The Restless Urban Landscape

CONTRIBUTORS

Robert A. Beauregard is a professor of planning at the University of Pittsburgh.

M. Christine Boyer is a professor of architectural history and urbanism at Princeton University.

Darrel Crilley is a graduate student in the Department of Geography and Earth Science, Queen Mary College, London, England.

Anthony D. King is a professor of art history and sociology at the State University of New York at Binghampton.

Paul L. Knox is a professor of urban affairs and planning at Virginia Polytechnic Institute and State University.

Scott Lash lectures in sociology at Lancaster University in England.

Robin M. Law is a graduate student in the Department of Geography at the University of Southern California.

David Ley is a professor of geography at the University of British Columbia.

John Logan is a professor of sociology at the State University of New York at Albany.

Caroline Mills lectures in geography at the College of St. Paul and St. Mary, England.

Jennifer R. Wolch is a professor of geography at the University of Southern California.

The Restless Urban Landscape

Paul L. Knox

EDITOR

Prentice Hall
Englewood Cliffs, New Jersey 07632

Library of Congress Cataloging-in-Publication Data

The Restless urban landscape / Paul L. Knox, editor.
p. cm.
Includes bibliographical references and index.
ISBN 0-13-755414-1
1. Cities and towns. 2. Urban geography. I. Knox, Paul L.
HT151.R427 1993
307.76--dc20 92-19473
CIP

Acquisitions Editor: Ray Henderson
Cover Concept: Anne-Lise Knox
Cover Designer: DeLuca Design
Prepress Buyer: Paula Massenaro
Manufacturing Buyer: Lori Bulwin

A Simon & Schuster Company
Englewood Cliffs, New Jersey 07632

Printed in the United States of America

10 9 8 7 6 5 4 3

ISBN 0-13-755414-1

Prentice-Hall International (UK) Limited, *London*
Prentice-Hall of Australia Pty. Limited, *Sydney*
Prentice-Hall Canada Inc., *Toronto*
Prentice-Hall Hispanoamericana, S.A., *Mexico*
Prentice-Hall of India Private Limited, *New Delhi*
Prentice-Hall of Japan, Inc., *Tokyo*
Simon & Schuster Asia Pte. Ltd., *Singapore*
Editora Prentice-Hall do Brasil, Ltda., *Rio de Janeiro*

CONTENTS

PREFACE

In November 1989 I found myself in what at that time was called Eastern Europe. I had been invited to give a series of lectures and seminars on the topic of urban change in the United States. As I sought to put the topic in perspective for audiences who were living through changes of almost unbelievable pace and proportion, I found myself arguing that American cities were also experiencing some significant changes, albeit less dramatic than those confronting the residents of Budapest, Warsaw, and East Berlin. The changes I had in mind were the emergence of a new metropolitan form, the development of unprecedented 'postsuburban' land-use patterns, the selective recentralization of economic activity, the socio-spatial polarization of central cities, and new architectural styles and modes of urban design. Set in the context of the evolving political economy of capitalist urbanization, I argued, a case can be made that these changes add up to the most pronounced restlessness in urban landscapes since the late nineteenth century, when street cars and elevators turned cities inside out and upside down.

As I reflected on all this, I realized that although I knew of a fair amount of recent research on various aspects of these changes, the material was not always very accessible. Hence the idea for an edited volume that would bring together some of the latest ideas and research on the topic. This book is the result. All of the chapters were written specifically for this book, though the reader will find that the writers have adopted different voices in their attempt to reach our collective audience. I have not attempted to edit these voices, since an important part of the exercise is to present to the reader a range of perspectives on various aspects of the restlessness urban landscape.

As always, an edited volume such as this is the product of the efforts of many people. Several of the contributors have recognized specific individuals in acknowledgements of their own; I would like to thank Melissa Chase, Janet Town-Tribble and, in particular, Krystal Wright, for help in preparing camera-ready material.

The Restless Urban Landscape

One

CAPITAL, MATERIAL CULTURE AND SOCIO-SPATIAL DIFFERENTIATION

Paul L. Knox

Virginia Polytechnic Institute and State University

American cities are not what they were a few years ago. Preliminary census returns for 1990 show that for the first time more than half of the nation's population lives in metropolitan areas of a million or more people. Among the 20 largest metropolitan areas, eight (Los Angeles, Washington, Dallas, Miami, Atlanta, Seattle, San Diego and Phoenix) grew by more than 20 percent between 1980 and 1990. But the changes experienced by cities go far beyond demographics. Cities *look* different. They have a decisively different form, structure and appearance. And, within the framework of newly developed and redeveloped tracts of urban land, beneath the exoskeleton of new architectural styles and redesigned cityscapes, there have emerged some important new cultural, social and political dynamics. Of course, cities are constantly changing and adapting. The built environment, in particular, bears witness to the 'restless formation and reformation of geographical landscapes' (Harvey, 1985, p. 150) in response to the imperatives and contradictions inherent to the dynamics of a capitalist economy and society. This restlessness has been particularly pronounced, however, since the mid-1970s. Epochal changes in the world economy have brought new needs, opportunities and tensions that have quickly been written into the landscapes of cities.

A new urban geography: packaged landscapes and stealth cities

The textbook geometry of sectors and zones has become increasingly difficult to discern in the landscapes, social ecologies, and bid-rent patterns of cities. The classic mosaic of central city neighborhoods has become blurred as the distinguishing features of class, race and family status have been overscored by lifestyle and cultural preferences. Long-established central city neighborhoods have either fallen into decay and social disorganization or have been 'reclaimed' by members of the bourgeoning new professional middle classes. Central Business Districts (CBDs) have experienced a selective recentralization of economic activity that has brought a renaissance of urbanity, a rash of speculative building and a sudden move toward conserving selected fragments and re-creating idealized tableaux of past development. Beyond the central city, suburban strips and subdivisions have been displaced as the conventional forms of new development by exurban corridors, office parks, business campuses, privately-planned residential communities, and outlying commercial centers big enough to be called 'edge cities.' Many of these, having grown up on greenfield sites near interstate junctions, straddle several administrative areas. As a result, they fail to show up on census charts or electoral maps and may even remain unnamed on street maps. They are 'stealth cities,' economically powerful but politicially invisible. They contain impressive concentrations of office jobs and hotel space and account for extremely high volumes of retail sales, but they have no parks, libraries or police forces.[1] Peirce Lewis described the outcome of these new forms of urban development as the 'galactic metropolis,' a new urban geography in which the individual elements 'seem to float in space; seen together, they resemble a galaxy of stars and planets, held together by mutual gravitational attraction, but with large empty areas between clusters' (1983, p. 35). The galactic metropolis is characterized by *packaged* landscapes and by landscapes of mixed densitites and unexpected juxtapositions of forms and functions. Meanwhile, new urban tissue throughout the metropolis has appeared in strikingly novel architectural styles.

So, just as it seemed that our theories and models might have captured the essential truths of urban geography, the transformation of cities themselves has made many of the models obsolete, forced a reevaluation of theory, and raised new issues that new models and revised theories must accommodate (Cooke, 1990a). From these circumstances there has emerged a widespread recognition of the importance of addressing the constitutive relations between class and space—what Soja (1980) termed the 'socio-spatial dialectic.' In this context, the built environment occupies an important position as both the product

[1] Tysons Corner, just beyond the Beltway around Washington, D.C., is the archetypal 'stealth city.' In the mid-1960s Tysons Corner was literally a rural corner of northern Virginia, marked only by the intersection of Interstate 66, the Washington Beltway and the access road to Dulles International Airport. Administratively, it is still rural, an unincorporated 6,000-acre area that contains 30,000 residents and over 75,000 jobs, all under the jurisdiction of Fairfax County but split between three different county supervisory districts and three county planning districts. Tysons Corner does not exist as a postal address: residents' mail must go to either Vienna or McLean. Within this 'stealth' framework in 1990 was the ninth largest concentration of commercial space in the United States, including more than 20 million square feet of office space, 3,000 hotel rooms, and parking for more than 80,000 cars. Yet, while it had become the largest retail concentration on the East Coast with the exception of Manhattan, it had little of the apparatus of urban governance or civic affairs. It had a branch of Tiffany's but no public open space; an exclusive business club but no public forum; acres of parking lots but no transit stop; dozens of sportswear stores but no recreation centers, swimming pools or bicycle tracks. It was a place of great affluence but at the same time a focus of intense concern over traffic congestion, inflated land values, service provision and land-use conflicts. See Brown and Hickok, 1991.

of, and the mediator between, social relations. Shifting material landscapes, observes Sharon Zukin, 'bridge space and time; they also directly mediate economic power by the *conforming to* and *structuring* norms of market-driven *investment, production and consumption*' (1988a, p. 435, emphasis added).

In short, the built environment is a good place to begin in order to make sense of the new geographies that are being inscribed onto the old framework of urbanization. New forms of urban development, new architectural styles, and new cityscapes are important elements of these new geographies in their own right (Knox, 1991). More compelling, though, is their significance as a *text* that can be 'read' in order to reflect the imperatives of economic, social, cultural and political forces at particular times, to reveal the relationships between the social and spatial dimensions of urbanization, to interpret the ideological content of socially-created space and to suggest the conflicts, tensions and contradictions involved in the process of urban development.

The built environment, then, must be seen as simultaneously dependent and conditioning, outcome and mechanism of the dynamics of investment, production and consumption. Approaching the built environment in these terms presents a considerable challenge. It must be treated as part of the totality of urban change. Specific outcomes—new architectural styles, new forms of residential development, etc.—must not be abstracted from the broader sweep of socio-spatial change. At the same time, this broader sweep must be recognized as a discontinuous sequence. It is complex and differential, both spatially and temporally. It follows that changes in the built environment must be situated not only in terms of the structural transformation of economies and societies but also in terms of the behavior of particular agents of change and groups of individuals in particular localties at particular times.

This book represents an attempt to begin to meet this challenge and to contribute to a new urban geography that addresses new urban realities. In this chapter, recent transformations in patterns of investment, production and consumption are examined, and an overview is presented of some of the new dimensions of urbanism associated with these transformations. We begin with a review of trends in investment in the urban development industry, which leads to a review of the changes that have taken place in the organization of the production of the built environment and an exploration of Michael Ball's concept (1986) of 'structures of building provision' as applied to the role of the urban design professions and the role of the state.

Capital and the structures of building provision

When petroleum prices quadrupled in 1973, Western economies were already exhibiting the symptoms of an epochal change. Along with the deindustrialization of the labor force that accompanied the shift toward a service-oriented economy, economic growth in general had been slowing, rates of profit were falling, inflation was rising and concern was increasing over international monetary instability. In the wake of the system shock precipitated by the rise in energy costs, the corporate world had to develop new strategies in order to survive, while those companies that were successful, and others with 'petrodollars' to invest, had to seek profitable outlets in a world of financial destabilization, intense competition and stagnating markets. This situation has been described by Harvey (1978) as a crisis of 'overaccumulation.' It was characterized by overproduction, surplus labor (high rates of unemployment and underemployment), a surplus of capital in the form of idle productive capacity, and a surplus of money capital. According to Harvey, such a crisis can only be resolved through the temporary switching of investment from the 'primary circuit' of industrial production to the 'secondary circuit' of capital assets that eventually aid production (e.g. factories, offices, roads) or stimulate consumption (e.g., housing) or to the 'tertiary circuit' of projects such as education or technology research that eventually improve the quality or productivity of labor.

Speculative investment in the built environment, as the major component of the secondary circuit, can thus be seen to account for much of the property boom experienced in the larger cities of Europe, North America and Australia during the late 1970s and the 1980s, the boom that was an important component of the 'cowboy economy' and growth culture of the 1980s[2] and that animated much of the restlessness in urban landscapes with which this book is concerned. As Bob Beauregard shows in Chapter 3, it is important to recognize that the spatial and sectoral shifts in capital investment, disinvestment and reinvestment that underpin this restlessness are not always neatly sequential. Moreover, it is clear, as Harvey himself recognized, that the built environment is given to the same vulnerability to overaccumulation as industrial production. Cities can become overbuilt with office blocks, shopping malls, business parks and planned communities, as many American cities had by 1990.[3] But, between the building booms of the 1970s and 1980s and the stalled growth of the early 1990s, a number of significant changes had occurred in the organization of real estate and urban development industries.

[2] Lending by commercial banks in the US against property increased from 29 percent of their assets in 1980 to 37 percent by 1989. See *The Economist* 317, No. 7679, November 1990, p. 19.

[3] In 1990, downtown office vacancy rates in Austin, Dallas, Phoenix, San Antonio, New Haven, Memphis, Miami, Houston and Denver were in excess of 20 percent. It is likely to take between 5 and 10 years to fill this surplus space.

As in other sectors, economic transformation in these industries has been characterized by the concentration and centralization of activity. The former involves the elimination of smaller, less profitable firms, partly through competition and partly through mergers and takeovers. Centralization involves the merging of the resultant large enterprises from different spheres of economic activity to form large, diversified corporations that are often transnational in their operations, having established overseas subsidiaries, taken over foreign competitors or bought into profitable foreign enterprises. As these new corporate structures have evolved in response to the contingencies of an overaccumulation crisis, they have also rationalized their operations in a variety of ways, redeploying their resources, reorganizing their production systems and revising their product mix.

The U.S. shopping center industry provides a good example of consolidation and centralization within the development industry.[4] During the 1970s, mergers and acquisitions led to the concentration of the major retailing assets within regional shopping malls—the big department stores—under the ownership of a few corporations (including Allied Stores, Associated Dry Goods, Federated Department Stores, R.H. May & Co. and May Department Stores). Meanwhile, ownership of the malls themselves became consolidated in the hands of a few big companies (including the Hahn Company, the Rouse Company, Melvin Simon & Associates and the Edward J. DeBartolo Corporation); and before long the largest retailing organizations had established their own development companies. Sears, for example, established the Homart Development Company; May established May Centers; and Federated established Federated Stores Realty. During the 1980s, the pace of mergers and acquisitions increased, partly because underutilized real estate assets made department store companies attractive targets for leveraged buyouts and partly as a result of development companies acquiring department store companies, as in the Canadian-based Campeau Corporation's acquisition of Federated (including the Bloomingdales, Burdines, Lazarus and Rich's chains) and Allied (including the Bon Marché, Jordan Marsh, Maas Bros. and Stern's chains). Centralization also increased as real estate became popular with institutional investors such as insurance companies and pension funds and as deregulation of financial markets removed many of the institutional barriers that had separated investments in residential development from those in non-residential development and in stocks and bonds (Downs, 1985). In particular, the 'securitization' of real estate assets meant that property financing came to be linked more directly to broader capital markets and, indeed, to international capital markets, as John Logan shows in Chapter 2.

By 1990, the shopping center industry was both highly concentrated and centralized, along with other branches of real estate and urban development:

[4] The following account draws on Mangan (1990); see also Hallsworth (1991).

It is clear that an 'hourglass' structure is emerging within the real estate industry. At the large top end of the hourglass are the large investors and developers like Prudential Insurance, Equitable Life, Metropolitan Life, Trammel Crow, Maguire Thomas, JMB Realty, Olympia and York, and similar firms. Virtually no real estate players occupy the thin middle of the hourglass, as there are few competitive medium-size firms. All the other real estate players are at the bottom and range from one-man brokerage firms to syndicators and small developers.

Ori, 1990, p. 96

The top 25 percent of diversified development companies (i.e., those engaged in developing offices, shopping centers and hotels) had increased their market share to almost 75 percent, while in the traditionally very decentralized house-building industry (there are over 110, 000 firms nationally) the top 0.09 percent of the firms had increased their market share to over 15 percent (Schwanke, 1989). As a result of this concentration and centralization, the *scale* of investment in the built envrironment has changed significantly. The involvement of congolmerate corporations and large financial institutions has made it possible to put together large-scale, highly visible projects. The typical mixed-use development (MXD) or multi-use development (MUD)[5] of the late 1980s required an investment of about $200 million, spread over 10 years or so, before any return could be expected. This, in turn, has meant that many of the larger development companies have moved into asset and property management, financing and brokerage in order to fully utilize the skilled professional work force that has to be assembled in order to put together such projects and follow them through. The scale of corporate organization, meanwhile, has meant that urban development is no longer predominantly a local activity (Leyshon, Thrift and Daniels, 1990; Logan, 1990; Strassman, 1988); indeed, it has an important international dimension, as Anthony King shows in Chapter 4. Olympia and York, the company whose activities are examined in detail by Darrel Crilley in Chapter 6, have assets of $31.2 billion and annual revenues of $8.7 billion (1990) and more than 100 million square feet of office space in cities across North America, including trophy structures such as the World Financial Center (formerly Battery Park City) and the Park Avenue Atrium in New York, Exchange Place (Boston), Yerba Buena Gardens (San Francisco) and First Canadian Place (Toronto). They also have extensive overseas operations, including the $6.6 billion Canary Wharf in London; they have signed a letter of intent to build an 80-story office block in Moscow and are

[5] A mixed-use development is defined as one involving several mutually supporting revenue-producing uses (e.g., retail, hotel, office, residential, entertainment) in projects that involve some integration of project components under a coherent plan. Multi-use developments are those where project components are not integrated under such a plan or where only two uses are involved.

bidding to build Tokyo Bay, a 15-million-square-foot office complex near downtown Tokyo. They have interests in other development companies, including Campeau, Landmark, Stanhope Properties, Trizec and Rouse, and have diversified into newsprint (Abitibi-Price), oil (Gulf Canada), financial services (Trilon), food and alcoholic beverages (Allied-Lyons), utilities (GW, Consumers Gas of Toronto) and railroads (Santa Fe).

The significance of these giant corporations lies in their potential for removing much of the debate and control over patterns of development from local arenas of municipal government, from the influence of local 'growth machine' coalitions and from the voice of neighborhood and environmental groups. Although large national and international development companies often work with local partners in joint ventures in order to exploit local networks of key contacts, design professionals and construction companies, the larger, stronger partners tend to learn quickly, setting up local subsidiaries or simply gaining the experience to take on additional projects on a full ownership basis.[6] Even with joint ventures, the use of non-local sources of development financing and the degree of non-local control over project design and management inevitably changes the dynamics of local politics, as John Logan shows in Chapter 2. Not least among these changes are the tendencies for urban land to be treated increasingly as a pure financial asset and for managers in both the private and public sectors to pursue more intensively the rationality of rent- and profit-maximizing and exchange value at the expense of other criteria, particularly use value (Haila, 1988).

Flexible strategies and lean production

While consolidation and centralization have been important aspects of economic transformation during the overaccumulation crisis, there have also been important changes in the organization of production. Here again, the changes exhibited in the development industry parallel changes that have occurred throughout Western economies. In broad terms, there has been a shift away from a reliance on economies of scale through the promotion of mass production and mass consumption (the so-called 'Fordist' approach to production) to an emphasis on 'lean' and 'flexible' production systems that are able to exploit new technologies such as robotics, computers and telecommunications systems. The objectives are to be able to target those sections of the market that are most profitable, to eliminate waste materials and to minimize labor costs and the costs of

[6] For example, 68 percent of Japanese development companies' investment in new construction projects in the U.S. in 1989 was through joint ventures with local partners. In 1989 it was 87 percent. See Susanna McBee (1990).

making, storing and moving large quantities of products that do not have committed buyers. An outstanding example and pioneer of lean and flexible production systems is the Benetton Group of clothing manufacturing and retailing.[7]

Not all of the features of lean and flexible production sytems employed by manufacturers and retailers could apply directly to the development industry (which can claim in any case to have been pursuing such systems for longer than most industries, despite the appearance in the 1940s of mass-production residential subdivisions and in the 1960s of modular, factory-built housing). Nevertheless, the production and marketing principles illustrated so well by the Benetton Group can be seen increasingly in urban development. Product differentiation and niche marketing, in particular, can be seen throughout the development industry, while some products are themselves being designed in order to address the very flexibility that manufacturers and retailers require. As Harvey puts it,

> *The frenetic pursuit of the consumption dollars of the affluent has led to a much stronger emphasis on product differentiation . . . Producers have, as a consequence, begun to explore the realms of differentiated tastes and aesthetic preferences in ways that were not so necessary under a Fordist regime of standardised accumulation through mass production.*
>
> (1989b, pp. 273–4)

In the commercial sector, product differentiation has resulted in a variety of new formats for hotels: luxury/full service, executive conference resorts, extended stay (with kitchen and laundry facilities en suite), economy-only, and all-suite. Developments for retailing have similarly seen different formats for different market segments: upscale downtown gallerias, for example, and 'power centers' (community shopping centers

[7] The company has grown from a single small factory near Venice in 1965 and a single retail outlet in the Alpine town of Belluno in 1968 to a global organization with more than 5000 retail outlets in 60-plus countries and its own investment bank and financial services operations. It has done so by exploiting computers, new communications and transportation systems and robotics to the fullest possible extent. Designers create shirts and sweaters on CAD terminals. But their designs are produced only for orders in hand, allowing for the coordination of production and the purchase of raw materials—the Japanese *kanban*, or 'just-in-time' system. In factories, rollers linked to a central computer spread and cut layers of cloth in small batches according to the numbers and colors ordered by Benetton stores around the world; sweaters, gloves and scarves, knitted in volume in white yarn, are dyed in small batches by machines similarly programmed to respond to sales orders. Completed garments are warehoused briefly (by robots) and shipped out directly to individual stores to arrive on their shelves within ten days of being made. Sensitivity to demand, however, is the foundation of the Group's success. Niche marketing and product differentiation have been central to this sensitivity, which requires a high degree of flexibility in exploiting new product lines. Key stores patronized by trend-setting consumers (such as the store on the Rue Fauborg St Honoré in Paris) are monitored closely, and many Benetton stores' cash registers operate as point-of-sale terminals, so that immediate market input is available daily. Another notable feature of the Group's operations is the way that different market niches are exploited with the same basic products. In Italy, Benetton products are sold through seven different retail chains, each with an image and decor calculated to bring in a different sort of customer.

located near regional shopping malls and dominated by specialized discount outlets). Another significant new 'product line' for developers is the specialized mall: a 'medical' mall, for example, that is crafted to provide busy, affluent consumers with one-stop shopping that offers physicians, counselors, therapists, medical laboratories, pharmacies, outpatient facilities, fitness centers, health food stores and cafes. Developers of business and industrial parks, meanwhile, have recently begun to offer 'flex space': single-story structures with 'designer' frontages, loading docks at the rear, and interior space that can be used for offices, R & D labs, storage or manufacture, in any ratio. Old product lines can also be 'treated' in order to enhance flexibility within the market. Business and industrial parks have been repackaged as 'planned corporate environments' with built-in daycare facilities, fitness centers, jogging trails, restaurants and convenience stores, lavish interior decor and lush exterior landscaping. In the residential sector, both larger firms and local developers have 'repositioned' themselves away from single-family 'starter' homes to build more multifamily projects (that, like business parks, are packaged with services: in this case, security systems, concierge services, exercise facilities, bike trails, etc.) or more expensive homes for the 'move-up' market (where the basic product—single-family suburban homes—is differentiated by features such as dramatic master bedroom/bathroom suites, expensive-looking materials, and signature landscaping). Large, privately-planned communities also became popular with developers during the 1980s because, as the Urban Land Institute's 1989 review of real estate trends puts it: 'They permit developers more flexibility in design and product type . . . [and] enable developers to respond quickly to changing market demand' (Suchman, 1989, p. 38).

Structures of building provision: place entrepreneurs, the state and the design professions

As Michael Ball has pointed out (1986), the production of the built environment is not simply a function of the behavior of a core of central actors on a stage set by broader economic forces. To be sure, the reactions and interactions of landowners, investors, financiers, developers and builders in changing circumstances provide, as outlined above, an important basis for the plot. But the detailed outcomes, the full production, depend on sets of time- and place-specific social and economic relations involving other actors, including design professionals, construction workers, business leaders and community leaders, and consumers. In addition, the state—both local and national—must be recognized as an important agent in its own right and as a site of conflict or cooperation between other actors. Ball (1986) called these sets of relations the 'structures of building provision' and pointed out that each needs to be seen in terms of its specific linkages (functional, historical, political, social, and cultural) with the wider environment. In this context, the changing role of the state is particularly important, along with the changing dynamics of the design professions that are often the conduit for linkages within the structures of building provision.

The state

In terms of the state, the most important changes that have occurred over the past fifteen years or so are the decline of the welfare state and the shift to entrepreneurialism in urban governance. Although deepening economic recession after 1973 accentuated the vulnerability of more and more people, recession also made it difficult, both politically and economically, to finance the welfare programs that had expanded dramatically in the 1960s (Castells, 1989, pp. 239–59). Taxpayer 'revolts,' led by Proposition 13 in California and Proposition 2½ in Massachusetts in 1978, presaged electoral victories by politicians with an ideological stance based on the belief that the welfare state had not only generated unreasonably high levels of taxation, a bloated class of 'unproductive' government workers and disincentives for 'ordinary' people to work and save, but also that it may have fostered 'soft' attitudes toward problem groups in society. With the ascension of Ronald Reagan to the Presidency, the federal government began a major round of restructuring and retrenchment. Federal responsibilities were decentralized and many spheres of activity involving federal oversight were deregulated. Federal outlays on urban-related programs declined by a third, in real terms, between 1978 and 1984 (Glickman 1984), while many city governments, facing a fiscal crisis intensified by the withdrawal of federal funds, drastically reduced municipal workforces and curtailed social programs and public services (Kantor, 1988).

At the same time, city governments drew on a long-standing tradition of boosterism to foster a new civic culture of entrepreneurialism. David Harvey (1989a) has observed that this new culture has at its center the notion of public-private partnership, in which public resources and legal powers are joined with private interests in order to secure external funding or investment. It has fostered a speculative and piecemeal approach to the management of cities, with greater attention being given to specific projects in particular locations rather than to city-wide services or investments (in health care, education, or housing, for example). Spectacular local projects such as downtown shopping malls, festival market places, new stadia, theme parks, and conference centers are seen as having the greatest capacity to enchance property values and generate retail turnover and employment growth. If successful, they can cast a beneficial glow over the whole city. Even in the face of poor economic performance, they can be regarded as a kind of 'loss leader' that has the capacity to bolster the image of a city and pull in other forms of development (Fainstein, 1991).

As a result, there has been a rapid proliferation of spectacular set-piece projects. The best-known examples include Baltimore's Harbor Place, Riverwalk in New Orleans, Riverfront in Savannah, Quincy Market in Boston, Pioneer Square in Seattle, South Street Seaport in New York and Atlanta's package for the 1996 Olympic Games. Even some of the best-known, largest and most spectacular of these projects have encountered financial difficulties, however, partly because of the inherent high risks of such projects and partly because intense competition between cities for investment dollars has contributed to overbuilding. It is also worth noting that the culture of deregulatory corporatism and civic entrepreneurialism has been linked with an increase in the scale and intensity of corruption, some of the most striking examples being intimately involved in patterns of urban development (Dear, 1990). Under the tenure of former Secretary Samuel Pierce, the U. S. Department of Housing and Urban Development is thought to have lost over $4 billion in waste, fraud and influence peddling. Meanwhile, the deregulation of the savings and loan industry in the early 1980s allowed new investors to use federally-insured savings deposits to fund a variety of speculative projects, including office buildings, shopping malls, MXDs and resort complexes, as well as junk bonds and land deals. The failure of hundreds of these projects and the collapse of the companies involved means that during the 1990s US taxpayers will spend $400–500 billion in paying off guaranteed deposits.

Planning

It should not be surprising that, in a period when urban governance has shifted its emphasis to entrepreneurialism, urban planning should have lost its way. Not so long ago, the planning profession was carried forward by visionary ideals and a conviction that goodwill and technical competence could achieve lasting improvements in urban life. It did not take long, however, to discover that planning could not deliver the utopian goods. Disillusionment with physical planning and regulatory controls, expressed in books such as *The Federal Bulldozer*,[8] *The Evangelistic Bureaucrat*[9] and *The Death and Life of Great American Cities*,[10] was followed by a retreat from social planning as slowed economic growth made many sections of society uneasy with redistributive activities. 'We went from dreaming big plans to trying to find small ones that could support a constituency. Planners and planning had abandoned, and were abandoned by, the production and growth sectors of the society. They were left with social amelioration in

[8] Anderson, M. 1964. *The Federal Bulldozer.* New York: McGraw-Hill.

[9] Davies, J.G. 1972. *The Evangelistic Bureaucrat.* London: Tavistock.

[10] Jacobs, J. 1961. *The Death and Life of Great American Cities.* New York: Vintage.

a society that gave it low priority' (Sternlieb, 1987, p. 22). In the consequent search for professional identity and credibility, planning became fragmented, pragmatically tuned to economic and political constraints and oriented to stability rather than being committed to change through comprehensive plans. Planning practice became estranged from planning theory and divorced from any broad sense of the public interest (Beauregard, 1990; Brooks, 1988; Friedmann, 1987). It became increasingly geared to the needs of producers and the wants of consumers and less concerned with overarching notions of rationality or criteria of public good. The outcome has been a disorganized approach that has led to a collage of highly differentiated spaces and settings. As Christine Boyer observes, 'fragmented elements of the city whole are planned or redeveloped as autonomous elements, with little relationship to the whole and with direct concern only for adjacent elements. In other words fragments of the city are regulated by special district or contextual zoning, Historic Preservation Controls, TDRs [Transfer of Development Rights] off of historic structures, and even the dictates of an EIS [Environmental Impact Statement], all of which pay atttention to the artful fragment but say nothing about the city as a whole' (1987, p. 6). In Chapter 5, Boyer critically examines the notion that such fragments are all we can expect from the design professions in an economy constrained by chronic fiscal crisis.

Meanwhile, as city governments began to shift toward entrepreneurialism, city planners were drawn increasingly into the business of public-private partnerships. As Beauregard puts it:

> *Economic development is so highly valued by elected officials that planners, even if they were not to share this ideology of growth, would find it difficult, if not impossible, to oppose the state's complicity. The result has been a peculiar form of nonplanning in which planners participate in individual projects, often attempting to temper the most egregious negative externalities, while failing to place these projects in any broader framework of urban development . . .*
>
> *Even the schools train students in real-estate development, the cutting edge of planning education. Planning has become entrepreneurial and planners have become dealmakers rather than regulators.*
>
> Beauregard, 1989, pp. 387–8

The deals that planners make include the provision of major infrastructural elements and concessions over the nature, scale and location of development in return for donations of land or packages of amenities such as transit improvements, daycare facilities, streetscape improvements and school buildings (Alterman, 1990; Lassar, 1990). The single most important aspect of this change in planning practice, however, has been the way that planning has accommodated developers' need for flexibility through new approaches to land-use zoning. Once the cornerstone of city planning pratice, single-purpose ('Euclidian' [11]) zoning is now seen as wasteful, monotonous and overly rigid. Mixed-use zoning, critical for developers wanting to produce large set-piece projects, is now seen by planners and planning committees as a means of everything from enhancing a city's tax base to initiating urban revitalization and increasing ridership on public transit systems. In downtown areas, mixed-use zoning is being combined with incentive systems in which developers are granted additional height or density allowances in exchange for specified building features (e.g., elaborate façades or building tops) or facilities such as daycare centers, residential space and space for services that might help restore variety and vitality to downtown districts. In suburban jurisdictions, the solution to the rigidity of Euclidian zoning has been cluster zoning and Planned Unit Development (PUD) zoning, whereby regulations are applied to an entire parcel of land rather than to individual building lots. With PUD zoning, developers can calculate densities and profits on a projectwide basis, allowing the clustering of buildings to make room for open spaces (such as golf courses) or preserve attractive site features (such as ponds or old barns) and facilitating a mixture of residential and non-residential elements and a mixture of housing types that can be adjusted as sales dictate. For developers, PUD zoning offers economies of scale plus scope for product diversity and flexibility, all within a predictable regulatory framework. For planners, PUD zoning offers the prospect of high-tax-yield development with services and amenities provided at no cost to the taxpayer.

Architecture

Like planning, architecture has lost its way. Even though the dynamics of the profession have always been dependent upon avant gardism, there were for a long time (roughly between 1920 and 1970) certain canons of 'good' architecture to which most architects subscribed. Drawn from the wellspring of the Bauhaus movement and institutionalized by CIAM (Les Congrès Internationeaux d'Architecture Moderne), these canons were aimed at the creation of 'pure' and 'timeless' architecture. Expressed in aphorisms such as 'less is more,' 'form follows function' and 'buildings as machines for living in,' the disciples of modernist architecture shared a belief 'that their new architecture and their new concepts of urban planning were expressing not just a new aesthetic image but the very substance of new social conditions which they were helping to create' (Carter, 1979, p. 324). Architecture was to be an agent of redemption. Through industrialized production, modern materials and functional design, architecture (as distinct from mere

[11] Named not for Euclidian geometry, which would be eminently appropriate, but for the test case of *Euclid* v. *Ambler*.

building) could be produced inexpensively, become available to all, and thus improve the physical, social, moral and aesthetic condition of cities. As Le Corbusier declared in his famous concluding lines of *Toward a New Architecture* (1927), 'Architecture or revolution. Revolution can be avoided.'

But, just as planning was unable to deliver the utopian goods, so modern architecture failed the test of reality.[12] The democratic symbolism of modernist architecture was co-opted by big business and translated into the iconography of corporate power, respectabiity and anonymity (Knox, 1987). More importantly, the occupants of modernist offices and apartment blocks did not always cooperate with the notion of cities and buildings as machines (or with the unspoken corollary: people as ciphers/passive beneficiaries of timeless design). The consequent disillusionment with the accepted canons of design has been documented in books such as *Form Follows Fiasco.*[13] A new avant garde, seizing the opportunity, offered new, user-friendly canons. Drawing on books such as *Defensible Space*[14] and *Learning from Las Vegas*,[15] new, post-Modern canons emerged: stylistic pluralism, the emphasis of scenographic, decorative and contextual properties of the built environment, the use of double-coding (the combination of Modernist styling or materials with something else, usually historic or vernacular motifs) and the rejection of the social objectives and determinist claims of modern architecture (Jencks, 1984; 1986). These canons have been hotly contested, however, with the result that the profession has, like planning, become fragmented, suffering a crisis of confidence and retreating into a pragmatic, project-by-project approach (Frampton, 1991; Sorkin, 1991; Ghirardo, 1991).

Meanwhile, as with planning, the exigencies of epochal changes in the political economy have also had their effects. Architectural firms have had to become lean and flexible. Larger firms have increased their market share and have reorganized their work force into design teams or 'studios' that specialize in particular kinds of projects or project components. At the same time, architects have forfeited their traditional role as master builder, whereby clients supported their creative acts while requiring little consultation in the design process. Patronage has turned to clientage, with developers imposing tight specifications while allocating much of the design process to other professionals such as engineers, interior designers, landscape architects, project consultants and construction managers.

[12] Jencks (1977, p. 9) dated the end of modern architecture to the afternoon of July 15, 1972, when the prize-winning Pruitt-Igoe housing project in St Louis was dynamited by city authorities after an unhappy history of crime and vandalism.

[13] Blake, P. 1974 *Form Follows Fiasco.* Boston: Little, Brown.

[14] Newman, O. 1972 *Defensible Space.* New York: Macmillan.

[15] Venturi et al. 1977 *Learning From Las Vegas.* Cambridge, Mass.: MIT Press.

High on the list of developers' specifications is often the requirement that a project be distinctive, singular in form or at least attention-grabbing. Such qualities help to create a clearly identifiable image for the project's occupants, to offer their employees a distinctive image with which to identify themselves and, not least, to justify higher rents. The following extracts [16] from a *Business Week* feature provide a telling insight:

> *Says Houston developer Gerald D. Hines, . . . "Companies spend a lot of money in advertising their image. They can cut that back if they have an outstanding building that gains national attention." A successful building image can also translate into more rental dollars. "Good architecture really helped New York's Citicorp Center," says one observer. "It has a skin that caught everybody's imagination, shape at the top, and large public areas . . . It's not in a great location, but it gets one of the highest rents in Manhattan," he says.*
>
> *Able to rent the building at a premium that real estate people estimate at 20 percent or more, Citibank Chairman Walter B. Wriston never moved himself or the majority of bank offices into the new structure. The design of Houston's Pennzoil Place also paid off in rents that command a premium of three to four dollars per square foot, says developer Hines. And the building also helped Pennzoil to attract geologists and engineers. "In a competitive job market," he says, "people want to work for a company that represents more than just a commodity in space."*
>
> *Business Week*, October 4, 1982, p. 125

Such sentiments have meant that 'name' architects are greatly in demand, not only because their superstar status helps ensure faster leasing at higher rents but also because they are able to bring a great deal of publicity to a project. Peter Eisenman, a self-consciously superstar architect, has claimed that he has been commissioned because of his ability to deliver 'cover shot' buildings. [17] As a result, 'name' architects like Eisenman have media careers that are embedded in the venal insecurity of the editors and publishers of glossy magazines. 'Superstar architecture,' meanwhile, has emerged as one of the hallmarks of late twentieth-century urban form. Examples include the AT&T building (New York), Pennzoil Place (Houston) and PPG Place (Pittsburgh) by Burgee and Johnson; the World Trade Center (New York) and Canary Wharf Tower (London) by

[16] Quoted in Scuri, 1990, pp. 15-16.1

[17] Public lecture, Virginia Polytechnic Institute and State University, September 19, 1990. The commission in question was the Koizumi Sangyo Building in Tokyo. It duly appeared as the cover shot of *Architecture* magazine's September 1990 issue.

Cesar Pelli; and the Portland Public Service building and the Humana Building (Louisville) by Michael Graves.

The postmodern turn: commodity aesthetics and the city of signs

The emergence of post-Modern canons of architecture has been part of a wider reconfiguration of socio-cultural values that has proceeded in tandem with the reconfiguration of the political economy. Postmodernity is manifest in many aspects of life, from architecture, art, literature, film, music and commodity aesthetics to urban design and planning. Postmodernism is properly taken to refer to the particular philosphy that impels specific aspects of postmodernity. Thus Foster (1985) made the important distinction between postmodernisms of resistance (that seek to to deconstruct Modernity and undermine the status quo) and postmodernisms of reaction (that are more superficial responses to Modernity, seeking merely to displace one style, fashion or system of practices with another). The 'postmodern turn' reflected in the geography of Western cities since the 1970s is overwhelmingly the result of postmodernisms of reaction, displacing—or at least competing with—the universalistic, rationalistic, purposive, hierarchical, ascetic and machismo motifs of Modernity with playful, ephemeral, anarchical, spectacular, combinatorial and androgynous motifs (Table 1.1). There is, however, some scope for a postmodernism of resistance in the urban landscape, as David Ley and Caroline Mills suggest in Chapter 10.

In fiction, postmodernity is manifest in the novels of Ballard, Barth, Burroughs, Doctorow, Eco, Pynchon and Rushdie. In film, it is found in the likes of *Blue Velvet* (directed by David Lynch), *Body Heat* (Lawrence Kasdan) and *Blade Runner* (Ridley Scott). In architecture and urban design the best-known exponents of the postmodern are Venturi, Graves, Moore, Bofill, Rossi and Krier. In art it can be found in the works of Sandro Chia, Carlo Mariani, Eric Fischl and Ron Kitaj; in interior design it is closely associated with Memphis and Sunar furniture, WMF and Alessi metalware and Swid Powell ceramics; and on television it is epitomized by MTV videos (Collins, et al., 1987; Connor, 1989; Featherstone, 1988; Jencks, 1984; Cooke, 1990b). It is most ubiquitous, appropriately enough, in advertising. Benetton (the archexponents of flexible production systems), for example, have tapped the androgyny of postmodernity.

Table 1.1 Some Attributes of Modernity and Postmodernity

Modernity	Postmodernity
ascetic	anarchic
autonomous	androgynous
avant-garde	combinatorial
centered	complex
creative	cosmopolitan
exclusive	decorative
future-oriented	eclectic
hermetic	ephemeral
hierarchical	hedonistic
machismo	historicized
meta-theoretical	impulsive
narrative	ironic
paradigmatic	narcissistic
rationalistic	neo-vernacular
totalizing	parodic
universalistic	pastiche
utopian	playful
	populist
	self-indulgent
	spectacular

Harvey (1989b) argues that, while this reconfiguration had been under way for some time, it was not until the overaccumulation crisis of the 1970s that the relationship between art and society was sufficiently shaken to allow full expression to postmodernity. It should be emphasized, however, that the postmodern turn does not amount to an eclipse of Modernity any more than flexible production systems have eclipsed Fordist systems. Postmodernity is an increasingly important dimension of socio-cultural life that articulates with certain features of economic and social change in contributing to the socio-spatial dialectics of the city. It should also be acknowledged that postmodernity is inherently Janus-like. One face of postmodernity is exciting and liberating: we are invited to abandon the security blanket of Modernity—the consoling myth of human liberation through economic and scientific progress—and to live for the moment. The other face is grimly imprisoning: we are made dependent on the constant recycling of themes, robbed of our cultural capacity to address the future, and confronted with dystopian scenarios of increasing violence, pornography, corruption, and economic instability and polarization (Cooke, 1990a).[18]

Above all, however, postmodernity is pluralistic and consumption-oriented. It is, as Heller observes (1990, p. 8) 'a wave within which all kinds of movements, artistic, political and cultural, are possible.' What is particularly important in the present context is that this pluralism provides the basis for highly differentiated patterns of consumption. Quoting Heller again:

> *The spectre of "mass society" in which everyone likes the same, reads the same, practices the same, was a short intermezzo in Europe and North America. What has . . . emerged is not the standardization and unification of consumption but rather the enormous pluralization of tastes, practices, enjoyments and needs. The quantity of money available for spending continues to divide men and women, but so do the kinds and types of enjoyment, practices, pleasure which they seek . . . More importantly, the different patterns of consumption have become embedded in a variety of* ***lifestyles*** *. . .*
>
> (p. 10, emphasis added)

This differentiation makes for a socio-cultural environment in which the emphasis is not so much on ownership and consumption per se but on the possession of particular *combinations* of things and the *style* of consumption. In addition, the space-time compression of new communications technologies and the consequent globalization and homogenization of consumer culture (Featherstone, 1990) has fostered the perceived need for distinctiveness and identity: 'this society which eliminates geographical distance reproduces distance internally as spectacular separation' (Debord, 1983, para. 167). In this 'society of the spectacle' (Debord, 1990) where emphasis is on appearances, the symbolic properties of urban settings and material possessions assume unprecedented importance. There is, of course, another link here with the reconfiguration of the political economy. The increased importance of style and design in material culture is in large part a response to the overall increase in the availability of housing and consumer goods made possible by Fordism. Social distinctions, previously marked by the ownership of particular kinds of consumer goods, now have to be established via the symbolism of 'aestheticized' commodities. An important aspect of this is postmodernity's reliance on the past. History is a potent source of distinction, redolent of heritage and authenticity and

[18] Penley (1989), observing the postmodern dependency on recycling ideas and the affinity in postmodern film and literature with dystopian scenarios, notes the success of *The Terminator* and the consequent appearance of soundalike titles that play on the dark side of postmodernity: *The Re-Animator, the Eliminators, The Annihilators* and the hard-core *Sperminator*.

brimming with themes that can be reproduced, recycled or combined to resonate with the double coding, pastiche, irony and eclecticism of postmodernity.

Commodities as signs

'Surrounded by our things,' writes McCracken (1988, p. 124), 'we are constantly instructed in who we are and what we aspire to. Surrounded by our things, we are rooted in and virtually continuous with our past. Surrounded by our things, we are sheltered from the many forces that would deflect us into new concepts, practices and experiences.' Patterns of consumption are among the most powerful and pervasive processes within the socio-spatial dialectic. They are epigrammatic, able to carry symbolic meanings; they mold people's consciousness of place, help people to construct real places and connect the key realms of nature, social relations and meaning (Sack, 1988). Indeed, so powerful are they that the act of shopping itself—even mere window shopping—can confer social prestige, as long as the setting is appropriately stylish (Shields, 1989).

Much of this power is a function of the web of signification ascribed to particular commodities and groups of commodities and the settings—homes, shops, workplaces—in which they are stored and displayed.

> *People want their stuff to tell them who they are. They ask that inanimate objects serve as stand-ins for deeper qualities. Not just pretty flowers but a built-in serenity is taken to exist in a Laura Ashley wallpaper pattern. Not just style but the character of a person is presumed to be made manifest by a Ralph Lauren blazer. People disappear into their clothes. Their conversation becomes merely a part of the ambience of the restaurants they frequent.*
>
> Shames, 1989, p. 66

If we are to be able to 'read' the city sensitively, we must be attentive to this broader web of signification. Yet signification bears no straightforward relationship to the material world. Signs and symbols 'reflect and refract another reality. Social life is impregnated with signs which make it classifiable, intelligible and meaningful' (Eyles, 1987, p. 95). Each signifier, whether it is a house or a fountain pen, an automobile or a pair of sneakers, can be ascribed not only a denotative, surface-level meaning but also one or more second-level, connotative meanings. Within a particular socio-cultural milieux, certain signifiers are transformed—'cooked,' in the terminology of Levi-Strauss (1970)—to form the basis of a socially constructed 'reality': a particular way of seeing the world. Of particular importance here is the consumer *totemism* by which the social identities of different groups are established and reproduced through their identification with particular objects (Eyles and Evans, 1987). Meyrowitz (1990) notes that, in contrast to the 'democracy of goods' (whereby consumption was a means of social integration) associated with Fordism, consumption is now viewed increasingly as a means of asserting distinctiveness within a mass society. 'Paradoxically, . . . the products that appeal are those that offer "universal uniqueness"—that is, that promise everyone the possibility—through consumption of the product—of being unique, special, out of the ordinary, apart from the crowd' (p. 131). We must recognize, however, that what is bought can be sold and what is acquired can be disposed of, so that the signified can become signifier and signs can become 'wild' (Foster, 1989). Postmodernity, because of its anarchic, impulsive and parodic qualities, makes for particularly fluid and unstable relations between signifier and signified, image and reality. The net result is a densely encoded text that is volatile and fragile, 'knitting together diverse signification fragments charged with mythologies, plural meanings and many different values. The text is never fully controlled by its author[s]; instead it is itself finally "finished" or "produced" in its reading by others, inevitably leaving meaning surpluses behind and fulfilling meaning deficits when they arise' (Luke, 1990, p. 7).

Because meanings must therefore be 'advertised' (in the broadest sense of the word) in order to be shared, advertising (in its narrower sense) has become an important mechanism in the social construction of reality. In addition to stimulating wants, advertising has always had a role in teaching people how to dress, how to furnish a home, and how to signify status through ensembles of possessions. In the 1970s and 1980s, however, the emphasis in advertising strategies shifted away from simple iconology (products as embodiments of status, glamor, and so on) to narcissism (products as instruments of self-awareness and self-actualization) and totemism (products as emblems of group stylishness) (Leiss, Kline and Jhally, 1986). Carefully targeted to specific market segments via 'psychographics'[19] and artfully exploiting new media technologies and techniques, advertising is largely responsible for what Berman (1982, p. 64) describes as the 'seamless aestheticization of everyday life' and what Jameson (1984) calls the 'culture of the image,' in which the boundaries between the image and the real are increasingly transgressed.

[19] Based on consumer surveys linked to cluster and factor analyses of socio-economic data by zip code: see, for example, Weiss, 1988.

Social, demographic and cultural change

Coincidental with (and implicated in) the postmodern turn have been some important demographic and social shifts that demand attention because of their role in the social production of urban space. At the heart of these shifts is the confluence of two processes. The first is the demographic wave that began in 1947 and stopped rather abruptly in 1964 after the introduction of the contraceptive pill. This is the 'baby boom' generation, the early cohorts of which found themselves entering labor and housing markets just as the economy began to feel the worst effects of the overaccumulation crisis. The second is the occupational polarization resulting from sectoral shifts in the economy combined with the effects of corporate reorganization and redeployment, advances in robotics and automation and the increasing participation of women in the labor force. The net effect has been an increase in the number of higher-paid jobs (in producer services, high-tech manufacturing and the media), an increase in the number of low-paid jobs (in routine clerical positions, retail sales, fast-food operations, etc.), and a decrease in middle-income jobs (skilled blue-collar manufacturing).

More specifically, new class fractions and changing patterns of vulnerable and disadvantaged households have appeared. Among the vulnerable sub-groups to increase in number in the 1970s were single-parent families (+22.8%), the lone elderly (+11.3%), immigrants (+53.8%) and persons on work disability (+19.5%). At the same time, an increasing proportion of large families and single-parent families were multiply-disadvantaged,[20] even though the overall incidence of multiply-disadvantaged households decreased (Knox and Rohr-Zanker, 1989). The trends that have contributed to these changes have also contributed to shifts within the urban system, so that there was a great deal of variability among central cities in the changing composition of vulnerable and disadvantaged populations. In Detroit, for example, where there was an overall loss of population (-21% over the decade), recent immigrants increased by 125,000 (51%), black households increased by 88,000 (11%), female-headed families increased by 31,000 (31%), and the incidence of families below the poverty line increased by 37,000 (14%). In Phoenix, where there was a large overall gain in population (+35%), black households only increased by 9,000 (25%) and poor families by 19,000 (22%), but female-headed families increased by 12,000 (43%) and recent immigrants increased by 34,000 (61%). In Los Angeles, where there was a more modest growth in population (+5.4% over the decade), black households declined in number by 10,000 (-2%), while there was significant growth in the number of female-headed families (33,000: 23%), poor families

[20] Experiencing two or more of the following: poverty-level incomes, unemployment, no telephone, incomplete kitchen or plumbing facilities and crowded living conditions.

(111,000: 23%) and recent-immigrant households (558,000: 57%) (Knox, 1990). These trends are discussed in Chapter 7, where Robin Law and Jennifer Wolch examine the implications of economic, demographic and social restructuring in terms of processes of social reproduction in contemporary cities.

Meanwhile, higher-income earners had emerged in occupations that had only a weakly established social status. Following Bourdieu (1984), they can be categorized as a 'new bourgeoisie' and a 'new petite bourgeoisie.' Thus, whereas the established bourgeoisie consists principally of intellectuals and industrial and commercial employers, the new bourgeoisie consists of members of the professions, public administrators, scientists, and private-sector executives, especially those involved in non-material products—economists, financial analysts, management consultants, personnel experts, designers, marketing experts, purchasers and so on. And, whereas the established petite bourgeoisie consists of craftspersons, small shopkeepers, office workers, and the like, the new petite bourgeoisie consists of junior commercial executives, engineers and skilled technicians, medical and social service personnel and, in particular, people directly involved in cultural production—authors, editors, radio and TV producers and presenters, magazine journalists and so on. As Table 1.2 shows, such occupations expanded dramatically in the 1970s, particularly in the larger metropolitan areas, where overall employment growth was relatively modest.

The new materialism

With the intersection of the maturing baby boom generation and emergent new class fractions, the cultural landscape has changed significantly, bringing a much greater emphasis on materialism and style only a few years after confident predictions of a pronounced societal trend of declining materialism (Inglehart, 1971; Bell, 1973). Heller (1990) suggests that the baby boomers, whose formative experience was the postwar economic boom, provided the main basis for such predictions. Their rebellion against the apparent complacency of industrial progress and affluence was channeled into a counter-cultural movement with a collectivist approach to the exploration of freedom and self-realization through a variety of movements, including radical politics, drugs, and sexual liberation. The high water mark of this 'alienation generation' was 1968, the year of sit-ins, student-worker alliances, protest marches, general strikes, civil disorder and riots. The failure of these events, observes Heller, produced the 'postmodern generation,' characterized by materialism and an unlimited pluralism. David Ley summarizes the transition as follows:

> *The post-modern project is the project of the new cultural class, representatives of the arts and the soft professions who came to political awareness in the 1960s and were receptive to the oppositional ideas of the counter-culture. But through the 1970s and 1980s hippies have all too readily become yuppies, as the subjective philosophies of phenomenology and existentialism which opposed the impersonality of modernism in the 1960s and redeemed the individual have been directed inwards, and the celebration of meaning has often shifted subtly to the celebration of meaning of the self.*
>
> (1986, p. 28)

Very quickly, self-awareness became a commodity as well as a state of mind. New product lines (and, significantly, magazines like *Self*) emerged to cater to this self-awareness and, as we have seen, advertising strategies began to exploit narcissism and totemism. Meanwhile, the maturing baby-boomers of the postmodern generation found themselves flooding the labor and housing markets just as the economy was experiencing the worst recession since the 1930s. Wages stood still while house prices ballooned. In 1973, 'the last really good year for the middle class' (Butler, 1989, p. 77), 'the average 30-year-old man could meet mortgage payments on a median-priced house with about a fifth of his income. By 1986, the same home took twice as much of his income. In the same years, the real median income of all families headed by someone under 30 fell by 26 percent.' Katy Butler offers this personal testament to the apostasy of the postmodern generation:

> *In our 20s, my friends and I hardly cared. We finished college (paid for primarily by our parents), ate tofu, and hung Indian bedspreads in rented apartments. We were young; it was a lark. We scorned consumerism. But in our 30s, as we married or got sick of having apartments sold out from under us, we wanted nice things, we wanted houses.*
>
> (p. 77)

Yet economic cirumstances did not permit a smooth transition to materialism, even for the college-educated middle classes:

> *My friends dressed and ate well, but despite our expensive educations, most had only one or two elements of the dream we had all laughed at in our 20s and now could not attain. We had to choose between kids, houses and time. Those with new cars had no houses; those with houses, no children; those with children, no houses.*

> *A few lucky supercouples—a lawyer married to a doctor, say, with an annual combined income of more than $100,000—had everything but time.*
>
> (p. 74)

Unable to fulfill the American Dream, the postmodern generation saved less, borrowed more, deferred parenthood, comforted itself with the luxuries that were marketed and written up in glossy magazines as symbols of style and distinctiveness and generally surrendered to the hedonism of lives infused with extravagant details: sun-dried tomatoes, imported mineral water, coarse-grain mustard, blackened redfish, and fresh-cut flowers on the table; clothes by Ralph Lauren, Perry Ellis, Issey Miyake and Liz Claiborne; Krups coffee makers, Braun juicers, Kitchen Aid appliances, Rolex watches, Nikon cameras, $50 haircuts, $200 shoes; Cancún in January; a cottage by the sea in July; Beaujolais Nouveau in November. The signs—literally—of this new materialism are everywhere. Bumper stickers advising SHOP TILL YOU DROP or boasting WHO DIES WITH THE MOST TOYS WINS; sweatshirts announcing DEAR SANTA: I WANT IT ALL; and college dorms decorated with posters depicting expensive cars, sushi arrangements, a bottle of champagne popping its cork.

The ambient culture of stylish materialism that developed in the 1980s also brought new patterns of social behavior. Barbara Ehrenreich (1989) argues that a cornerstone of the 'yuppie strategy' has been the determination of both men and women to find proven wage-earners as potential marriage partners. The result, she suggests, has been the consolidation of an androgynous caste- and guild-like class fraction characterized by the high educational status of both men and women and by very high household incomes. The importance of marrying suitably qualified partners has, in turn, intensified the potency of material signifiers:

> *. . . since bank accounts and resumés are not visible attributes, a myriad of other cues were required to sort the good prospects from the losers. Upscale spending patterns created the cultural space in which the financially well matched could find each other—far from the burger eaters and Bud drinkers and those unfortunate enough to wear unnatural fibers.*
>
> Ehrenreich, 1989, p. 229

Within the new urban geography, this new cultural space has come to be situated in upscale restaurants, department stores, condominiums, apartment blocks and mixed-use developments, all finished in expensive materials and equipped with extravagant details to complement materialistic lifestyles. These top-of-the-line products from the development industry are carefully pitched to the tastes of the professional middle classes who have become the avatars of the new culture of aestheticized consumption that has diffused throughout the entire middle class and beyond (Featherstone, 1991).

Table 1.2 Occupational Change in Selected U.S. Metropolitan Areas
Percent Employment Change 1970–1980

	United States	New York	Los Angeles	Chicago	Philadelphia	San Francisco	Washington	Dallas-Fort Worth
New Bourgeoisie (selected occupations)								
Accountants and auditors	55.8	2.2	26.4	29.6	22.2	53.8	44.2	119.6
Architects	98.3	17.2	81.0	61.9	39.6	79.9	53.6	156.9
Lawyers	83.0	25.0	104.7	78.0	95.1	144.3	200.9	159.7
Scientists	37.5	2.5	15.5	11.8	3.1	34.9	22.0	227.4
Social scientists and urban planners	97.4	45.5	60.0	67.8	64.9	90.7	54.3	217.1
New Petite Bourgeoisie (selected occupations)								
Counselors (vocational and educational)	69.0	28.0	101.0	46.8	92.8	48.8	60.3	213.2
Dieticians	54.0	27.6	95.9	68.0	32.2	73.1	225.7	187.4
Health technologists and technicians	81.2	113.8	206.8	197.2	195.5	188.6	260.6	536.3
Therapists	148.7	64.4	106.1	123.9	161.5	128.7	235.0	302.8
Writers, artists and entertainers	47.5	32.1	68.6	37.0	46.1	88.3	71.8	177.2
Designers	44.3	63.4	111.8	131.6	120.9	197.6	333.1	341.0
Painters, craft artists, etc.	75.3	−5.0	26.4	92.9	18.5	72.4	61.2	115.8
Photographers	38.2	8.4	33.8	12.0	10.6	34.2	35.2	127.0
Editors/reporters	34.6	21.8	45.3	12.0	30.8	61.7	34.6	116.6
Public relations specialists	47.1	6.7	31.3	31.0	25.2	53.9	219.3	128.6
Total Employed	***27.5***	***−15.2***	***22.8***	***13.6***	***5.9***	***25.7***	***29.9***	***123.7***

Source of data: U.S. Census of Population and Housing

The habitus of the middle classes: landscapes of conspicuous consumption

The values, cognitive structures and orienting practices of individual class fractions represent what Bourdieu (1984) has conceptualized as the *habitus*. It is a collective perceptual and evaluative schema that derives from its members' everyday experience and operates at a subconscious level, through commonplace daily practices, dress codes, use of language, comportment and patterns of consumption. The result is a distinctive pattern in which 'each dimension of lifestyle symbolizes with others' (p. 173). Each class fraction establishes, sustains and extends its habitus through the appropriation of 'symbolic capital': luxury goods attesting to the taste and distinction of the owner. That symbolic capital is vulnerable to shifts in avant garde taste, the availability of new product lines and new styles only makes it more useful as a measure of distinction. But the habitus of dominant class fractions is inevitably undermined by the popularization of goods and practices that were formerly exclusive, so that the struggle for distinction is continuous. As a result, the signs of distinction have constantly to be shuffled, inverted or displaced. New languages of taste have to be mastered; kitsch has to be consecrated as aesthetic; and the once-fashionable has to be condemned as kitsch.

In this context, as Bourdieu observes, the 'cultural capital' available to a class fraction is critical. Cultural capital is not simply equivalent to 'cultural literacy' but is, rather, a product of competence in the symbolic meaning of particular cultural artefacts. Such competence comes easily to the new bourgeoisie and new petite bourgeoisie, suffused as they are with designers, marketing experts and people directly involved in cultural production. It is central, for example, to the historic preservation described by Rowntree and Conkey (1980), who demonstrate how the sponsorship of certain components of the built environment assisted a specifc group to establish a cultural bridgehead 'through the creation of shared symbolic structures that validate, if not actually define, social claims to space and time' (p. 459).

This brings us to the question of how the habitus of the middle classes—the bourgeoisie and petite bourgeoisie, both old and new—are reflected in the new urban geography and to the deeper question of how urban landscapes constitute and reconstitute the lifeworld and the habitus of particular class fractions (Soja, 1992; Sorkin, 1992; Zukin, 1991). It must be sufficient for the moment to point to certain dimensions of the emergent geography of American cities that are of particular significance in this context. Among these is the role of consumption as the means by which people give meaning to their lives and the role of shopping as the framework around which increasing numbers of people structure their daily lives. Such phenomena translate into the urban landscape in a number of ways. Thus, for example, because style and distinction are so important to consumption (and even to the practice of shopping), the retail landscape has polarized:

> *Department stores, for example, faced the choice of specializing in one end of the class spectrum or another—or else going out of business. Undifferentiated chains, like Korvette's and Gimbel's, which had aimed at both blue- and white-collar middle-income consumers, were forced to close, while Sears and J.C. Penney anxiously tried to 'reposition' themselves to survive in the ever more deeply segmented market. The stores and chains that prospered were the ones that learned to specialize in one extreme of wealth or the other: Bloomingdale's and Neiman-Marcus for the upscale; K-Mart and Woolco for those constrained by poverty or thrift.*
>
> Ehrenreich, 1989, p. 228

Similarly, shopping malls are increasingly differentiated, with layout, interior design and iconography all carefully tuned to market niches (Crawford, 1992; Gottdiener, 1986; Shields, 1989).

Another important dimension of the emergent geography of American cities is the way in which the different combinations of money capital and cultural capital associated with particular class fractions give rise to preferences for different architectural styles, residential milieux and work environments. 'As the various social classes or groupings vie with each other for housing—or are catered to (or manipulated) by speculator-developers—their competition for urban space is manifested in intraurban migration and in changing residential differentiation' (King, 1989, pp. 874–5). Gentrification, loft living, the appropriation of areas of historic preservation and private, master-planned communities have already been nominated as being pivotal to the spatial practices of the professional middle classes (Dorst, 1989; Eyles and Evans, 1987; Knox, 1991; Mills, 1988; Smith 1987, 1992; Zukin, 1988b). Many of their employers, meanwhile—particularly advertising agencies, consultants, lobbyists, financial services, media services and the like—have been involved in a second dimension of intraurban migration, renting or building settings appropriate to their own corporate image and to the aesthetic preferences of key employees. Postmodern urban landscapes and the ecology of postmodern spaces are examined in some detail in Chapter 8.

The city of the excluded

If consumption has become the means by which people give existential meaning to their lives, it follows that low-income households do not exist! Unable to participate in the existentialism of the marketplace, they have become invisible to retailers, advertisers, developers and, indeed, the rest of society. At best they can manage a hollow emulation of the styles of the middle classes: slightly dated but cheap and awkward copies of clothes, homes and appliances—object-lessons in the price of middle-class failure. At worst, they are left with an unhealthy diet, shoddy goods and unsanitary homes. Deutsche and Ryan (1988), discussing the displaced residents of New York's Lower East Side, note that:

> *like the existence of the Palestinian people, the existence of the original residents of the Lower East Side is in the eye of the beholder. There were, in fact, over 150,000 people living in the area, thirty-seven percent Hispanic and eleven percent black . . . . The fact that forty percent of the population lives in official poverty might account for their high rate of invisibility.*
>
> (p. 103)

Andrew Mair (1986) has argued that the nature of the post-industrial city, with its emphasis on materialism, demands the invisibility—through displacement, removal, exclusion or segregation—of the homeless and the poor. Although urban space is produced and sold in discrete parcels, it is *marketed* in large packages: 'when an evening's entertainment, or a house, is bought, the neighborhood, and perhaps the whole central-city area, is consumed' (Mair, 1986, p. 363). The poor and the homeless represent potent negative externalities, not only endangering the exchange value of homes but also threatening the ability of upscale settings to deliver style, distinction and exlusivity, to insulate the middle classes from the anxiety induced by contact with the poor and to legitimize and reproduce the ideology of consumerism.

The exclusion and segregation of the poor is of course a well-worn theme in urban geography. Thus 'the existence in 1980s Los Angeles of $11 million condominium apartments, sold with a complimentary Rolls Royce, and, 50,000 homeless wandering the beaches of the Californian dream,' as Castells observes (1989, p. 224) 'is but an extreme manifestation of an old urban phenomenon, probably aggravated in the 1980s by the removal of the welfare safety net in the wake of neoconservative public policies.' It is clear, however, that economic restructuring and occupational polarization, combined with social and demographic change, have initiated cruelly widening inequalities and heightened estrangement along lines of class and race: 'The differential reassignment of labor in the process of simultaneous growth and decline results in a sharply stratified, segmented social structure that differentiates between upgraded labor, downgraded labor and excluded people.' The latter 'share an excluded space that is highly fragmented, mainly in ethnic terms, . . . reservations for displaced labor, barely maintained on welfare' (Castells, 1989, pp. 225 and 227). The new poor, in short, also have a place in the new urban geography (Knox, 1990).

Most striking among the landscapes of the excluded are 'impacted ghettos' (Hughes, 1990), spatially isolated concentrations of the very poor, usually, though not always, predominantly black and often drained of community leaders and containing very high proportions of single-parent families struggling to survive in downgraded environments that also serve as refuges for the criminal segment of the informal economy (Wilson, 1987; Kasarda, 1990). Less visible, but more decisively excluded, are the 'landscapes of despair' (Dear and Wolch, 1987; Wallis, 1991) inhabited by the homeless: micro-spaces that range from the vest-pocket parks of downtown Los Angeles, the city square in San Francisco and the federal area of the District of Columbia to the anonymous alleyways of every big city. As we shall see in subsequent chapters, these spaces are just as much a part of the restlessness of contemporary urban landscapes as MXDs, master-planned communities and gentrified neighborhoods.

References

Alterman, R. 1990 Developer obligations for public services in the USA, pp. 162-74 in P. Healey and R. Nabarro (eds.) *Land and Property Development in a Changing Context.* Brookfield, Vermont: Gower Publishing.

Anderson, M. 1964 *The Federal Bulldozer.* New York: McGraw-Hill.

Ball, M. 1986 The built environment and the urban question, *Society and Space*, 4, 447–64.

Beauregard, R. 1989 Between Modernism and Postmodernism: the ambivalent position of U.S. planning, *Society and Space*, 7, 381–96.

Beauregard, R. 1990 Bringing the city back in, *Journal of the American Planning Association*, 56, 210–15.

Bell, D. 1973 *The coming post-industrial society: A venture in social forecasting.* New York: Basic Books.

Berman, M. 1982 All that is Solid Melts into Air. New York: Simon and Schuster.

Blake, P. 1974 *Form Follows Fiasco.* Boston: Little, Brown.

Bourdieu, P. 1984 *Distinction: A Social Critique of the Judgement of Taste.* London: Routledge & Kegan Paul.

Boyer, C. 1987 The return of the aesthetic to city planning: Future theory as departure from the past. Paper presented at Rutgers University Center for Urban Policy Research Conference on Planning Theory in the 1990s, Washington D.C. (mimeo).

Brooks, M. 1988 Four critical junctures in the history of the urban planning profession, *Journal of the American Planning Association*, 54, 241–8.

Brown, J. E. and Hickok, M. E. 1991 Can Tysons be Saved?, *Regardie's* magazine, Sept./Oct., 43-52.

Butler, K. 1989 Paté poverty: Downwardly mobile baby boomers lust after luxury, *Utne Reader*, September/October issue, 72–80.

Carter, E. 1979 Politics and architecture: An observer looks back at the 1930s, *Architectural Review*, November, 325–8.

Castells, M. 1989 *The Informational City.* Cambridge, Mass.: Blackwell.

Collins, M. et al. 1987 *The Post-Modern Object.* London: Academy Group.

Connor, S. 1989 *Postmodernist Culture.* Oxford: Blackwell.

Cooke, P. 1990a Modern Urban Theory in Question, *Transactions, Institute of British Geographers*, 15, 331-43.

Cooke, P. 1990b *Back to the Future: Modernity, Postmodernity and Locality.* London: Unwin Hyman.

Crawford, M. 1992 The World in a Shopping Mall, pp. 3–30 in M. Sorkin (ed.) *Variations on a Theme Park.* New York: Hill and Wang.

Davies, J.G. 1972 *The Evangelistic Bureaucrat.* London: Tavistock.

Dear, J. and Wolch, J. 1987 *Landscapes of Despair.* Princeton: Princeton University Press.

Dear, M. 1990 Editoral: Geographics of corruption, *Society and Space*, 8, 249–53.

Debord, G. 1983 *Society of the Spectacle.* Detroit: Red and Black Books.

Debord G. 1990 *Comments on The Society of the Spectacle.* New York: Verso.

Deutsche, R. and Ryan, C.G. 1988 The fine art of gentrification, *October*, 12, 91–111.

Dorst, J.D. 1989 *The Written Suburb.* Philadelphia: University of Pennsylvania Press.

Downs, A. 1985 *The Revolution in Real Estate Finance.* Washington D.C.: The Brookings Institution.

Ehrenreich, B. 1989 *Fear of Falling.* New York: Pantheon.

Eyles, J. 1987 Housing advertisements as signs: locality creation and meaning systems, *Geografiska Annaler*, 69B, 93–105.

Eyles, J. and Evans, M. 1987 Popular consciousness, moral ideology and locality, *Society and Space*, 5, 39–71.

Fainstein, S.S. 1991 Promoting economic development, *Journal of the American Planning Association*, 57, 22–33.

Featherstone, M. 1988 In pursuit of the postmodern: An introduction, *Theory, Culture and Society*, 5, 195–215.

Featherstone, M. 1990 Global culture. An introduction, *Theory, Culture and Society*, 7, 1–14.

Featherstone, M. 1991 *Consumer Culture and Postmodernism.* Newbury Park, CA: Sage.

Foster, H. (ed.) 1985 *Postmodern Culture.* London: Pluto Press.

Foster, H. 1989 The breakup of the sign in seventies' art, pp. 69–86 in J. Tagg (ed.), *The Cultural Politics of Postmodernism.* Binghampton, N.Y.: Department of Art and Art History, SUNY.

Frampton, K. 1991 Reflections on the autonomy of architecture: A critique of contemporary production, pp. 17–26 in D. Ghirardo (ed.), *Out of Site.* Seattle: Bay Press.

Friedman, J. 1987 *Planning in the Public Domain.* Princeton, N.J.: Princeton University Press.

Ghirardo, D. (ed.) 1991 *Out of Site.* Seattle: Bay Press.

Glickman, N.J. 1984 Economic policy and the cities. In search of Reagan's real economic policy. Paper presented to the Regional Science Association Meetings, Chicago. Cited in Castells, 1989.

Gottdiener, M. 1986 Recapturing the Center: A semiotic analysis of shopping malls, pp. 288–302 in M. Gottdiener and A. Lagopoulos (eds.), *The City and the Sign.* New York: Columbia University Press.

Haila, A. 1988 Land as a financial asset: The theory of urban rent as a mirror of economic transformation, *Antipode,* 20, 79–101.

Hallsworth, A.G. 1991 The Campeau takeovers—the arbitrage economy in action, *Environment and Planning A*, 23, 1217–24.

Harvey, D.W. 1978 The urban process under capitalism: A framework for analysis, *International Journal of Urban & Regional Research*, 2, 101–31.

Harvey, D.W. 1985 *The Urbanization of Capital.* Oxford: Blackwell.

Harvey, D.W. 1989a From managerialism to entrepreneurialism: The transformation in urban governance in late capitalism, *Geografiska Annaler*, 71B, 3–17.

Harvey, D.W. 1989b *The Condition of Postmodernity.* Oxford: Blackwell.

Heller, A. 1990 Existentialism, Alienation, Postmodernism: Cultural Movements as Vehicles of Change in Patterns of Everyday Life, pp. 1–13 in A. Milner, P. Thompson, and C. Worth (eds.), *Postmodern Conditions.* New York: Berg.

Hughes, M.A. 1990 Formation of the impacted ghetto. Evidence from large metropolitan areas, 1970–1980, *Urban Geography*, 11, 265–84.

Inglehart, R. 1971 The silent revolution in Europe: intergenerational change in post-industrial societies, *American Political Science Review*, 65, 911–1017.

Jacobs, J. 1961 *The Death and Life of Great American Cities.* New York: Vintage.

Jameson, F. 1984 Postmodernism, or the cultural logic of capitalism, *New Left Review*, 146, 53–92.

Jameson, F. 1985 Postmodernism and consumer society, pp. 111–25 in H. Foster (ed.), *Postmodern Culture.* London: Pluto Press.

Jencks, C. 1977 *The Language of Post-modern Architecture.* New York: Rizzoli.

Jencks, C. 1984 *Late-Modern Architecture.* New York: Rizzoli.

Jencks, C. 1986 *What is Postmodernism?* New York: St Martin's.

Kantor, P.B., with S. David 1988 *The Dependent City.* Glenview, Ill.: Scott, Foresman.

Kasarda, J. 1990 Structural factors affecting the location and timing of urban underclass growth, *Urban Geography,* 11, 234–64.

King, R.J. 1989 Capital switching and the role of ground rent: Switching between circuits, switching between submarkets, and social change, *Environment & Planning A,* 21, 853–880.

Knox, P.L. 1987 The social production of the built environment: architects, architecture and the post-Modern city, *Progress in Human Geography,* 11, 354–378.

Knox, P.L. 1990 The new poor and a new urban geography, *Urban Geography,* 11, 213–6.

Knox., P.L. 1991 The Restless Urban Landscape: Economic and Socio-Cultural Change and the Transformation of Washington D.C., *Annals, Association of American Geographers,* 91, 181–209.

Knox, P.L. and Rohr-Zanker, R. 1989 *Economic change, demographic change and the composition and distribution of disadvantaged households in the United States.* Washington D.C.: U.S. Department of Commerce, Economic Development Adminstration.

Lassar, T.J. 1990 *City Deal Making.* Washington D.C.: Urban Land Institute.

Leiss, W., Kline, S. and S. Jhally 1986 *Social Communication in Advertising.* New York: Methuen.

Levi-Strauss, C. 1970 *The Raw and the Cooked.* London: Cape.

Lewis, P.F. 1983 The Galactic Metroplis, pp. 23–49 in R.H. Platt and G. Macinko (eds.), *Beyond the Urban Fringe.* Minneapolis: University of Minnesota Press.

Ley, D. 1985 Cultural/humanistic geography, *Progress in Human Geography,* 9, 415–23.

Ley, D. 1986 Modernism, postmodernism and the struggle for place. Paper presented to the seminar series 'The Power of Place,' Department of Geography, Syracuse University (mimeo).

Leyshon, A., Thrift, N. and P. Daniels 1990 The operational development and spatial expansion of large commercial development firms, pp. 60–97 in P. Healey and R. Nao (eds.), *Land and Property Development in a Changing Context.* Brookfield, VT: Gower Publishing.

Logan, J. 1990 From beyond city limits: local impacts of the globalization of real estate. Paper presented to Urban Affairs conference, Ohio State University (mimeo).

Luke, T. 1990 *Screens of Power*. Chicago: University of Illinois Press.

McBee, S. 1990 Japanese development deals in the United States, *Urban Land*, 49, 8, 2–5.

McCracken, G. 1988 *Culture and Consumption: New Approaches to the Symbolic Character of Consumer Goods and Activities*. Bloomington: Indidana University Press.

Mair, A. 1986 The homeless and the post-industrial city, *Political Geography Quarterly*, 5, 351–68.

Mangan, D. 1990 Consolidation of the shopping center industry, *Urban Land*, 49, 6, 30–31.

Meyrowitz, J. 1990 Commentary: On 'The consumer's world' by Robert Sack, *Annals, Assocation of American Geographers*, 80, 129–32.

Mills, C.A. 1988 Life on the upslope: The postmodern landscape of gentrification, *Society and Space*, 6, 169–89.

Newman, O. 1972 *Denfensible Space*. New York: Macmillan.

Ori, J.J. 1990 Real estate strategies for profitability in the 1990s, *Real Estate Review*, 20(1), 94–6.

Penley, C. 1990 Time travel, primal scene and the critical dystopia, pp. 33–50 in J. Tagg (ed.), *The Cultural Politics of Postmodernism*. Binghampton: Department of Art and Art History, SUNY.

Rowntree, L.G. and Conkey, M.W. 1980 Symbolism and the cultural landscape, *Annals, Association of American Geographers*, 70, 459–74.

Sack, R.D. 1988 The Consumer's World: Place as Context, *Annals, Association of American Geographers*, 78, 642–65.

Scuri, P. 1990 *Late Twentieth-Century Skyscrapers*. New York: Van Nostrand Reinhold.

Shames, L. 1989 What a long, strange (shopping) trip it's been. Looking back at the 1980s, *Utne Reader*, September/October, 66–71.

Shields, R. 1989 Social spatialization and the built environment: The West Edmonton Mall, *Society and Space*, 7, 147–64.

Smith, N. 1987 Of yuppies and housing: Gentrification, social restructuring and the urban dream, *Society and Space*, 5, 151–72.

Smith, N. 1992 New City, New Frontier: The Lower East Side as Wild, Wild West, pp. 61–93 in M. Sorkin (ed.), *Variations on a Theme Park*. New York: Hill and Wang.

Soja, E. 1980 The socio-spatial dialectic, *Annals, Association of American Geographers*, 70, 207–25.

Soja, E. 1992 Inside Exopolis. Scenes from Orange County, pp. 94–122 in M. Sorkin (ed.), *Variations on a Theme Park.* New York: Hill and Wang.

Sorkin, M. 1991 *The Exquisite Corpse.* New York: Verso.

Sorkin, M. (ed.) 1992 *Variations on a Theme Park. The New American City and the End of Public Space.* New York: Hill and Wang.

Sternlieb, G. 1987 Planning, American Style, *Society*, 25, 4, 21–3.

Strassman, W.P. 1988 The United States, pp. 22-58 in W.P. Strassman and J. Wells (eds.), *The Global Construction Industry.* London: Unwin Hyman.

Suchman, D.R. 1989 Housing and Community Development, pp. 28–37 in D. Schwanke (ed.), *Development Trends 1989*, Washington D.C.: Urban Land Institute.

Venturi, R. et al. 1966 *Learning from Las Vegas.* Cambridge, Mass.: MIT Press.

Wallis, B. (ed.) 1991 *If You Lived Here.* Seattle: Bay Press.

Weiss, M.J. 1988 *The Clustering of America.* New York: Harper & Row.

Wilson, W.J. 1987 *The Truly Disadvantaged: The Inner City, the Underclass, and Public Policy.* Chicago: University of Chicago Press.

Wood, J.S. 1988 Suburbanization of center city, *Geographical Review*, 78, 325–329.

Zukin, S. 1988a The postmodern debate over urban form, *Theory, Culture and Society*, 5, 431–46.

Zukin, S. 1988b *Loft Living, Culture and Capital in Urban Change.* New York: Radius Books.

Zukin, S. 1991 *Landscapes of Power. From Detroit to Disney World.* Berkeley: University of California Press.

Two

CYCLES AND TRENDS IN THE GLOBALIZATION OF REAL ESTATE

John Logan

State University of New York at Albany

Real estate has traditionally been a mostly local business. Real estate speculation has attracted a largely parochial crowd, investors of modest scale whose knowledge of the local market and connections to local officials gave them a competitive edge in this lucrative but risky business. This is not to say that capitalism's heavy hitters were ever *not* involved in real estate. Major corporations are necessarily large property owners, and property holdings have long been an important part of diversified investment portfolios. Nevertheless, real estate deals have tended to have more the flavor of Main Street than of Wall Street.

Real estate investors have traditionally been very active in local politics. Their business often requires government approvals and concessions: to build at a higher density than current law allows, to get sewers or sidewalks installed at public expense, to receive loan subsidies from municipal or county agencies. They have fueled the 'growth machines' that Molotch (1976) has identified in control of municipal politics in most cities. This is why Domhoff (1983, p. 197), in his analysis of 'who rules America,' is careful to distinguish 'between a national corporate community based in production and local growth machines based in land use.'

There are signs that this national-local distinction may be breaking down. This would not be surprising in a world where fortunes can be shifted across continents in an instant and where the futures of both small towns and big cities often seem to depend on investment decisions made in boardrooms on the other side of the world. This is an era when localities seem to be decisively penetrated by global trends. My concern here is with changes in the real estate industry, which is one economic sector among many. The local players, it seems, are being replaced with national and international conglomerates. There are development companies today that can operate simultaneously in Singapore and

Cincinnati, that can draw upon investment funds from Switzerland to build a hotel in San Francisco, that can raise a billion dollars in a single deal. How does this affect the process of urbanization and of political decision-making at the local level about development projects? Is the locality, on which American dreams of grassroots of democracy and home rule are based, fading into history, along with the general store and the milkman?

I have some doubts to express here. It is common now to find portayals of 'globalization' and 'economic restructuring' as inexorable forces following their own internal logic. They are appealed to as the cause of urban and regional changes beyond the power of cities to contain or resist (this point is further argued in Logan and Swanstrom, 1990). I wish to look more closely at this portrayal: in what ways are markets being restructured and what consequences do these changes have for social life?

Outside a relatively specialized business press, it is difficult to find empirical descriptions of the actual operation of real estate markets. Much is written about land-use patterns, gentrification and world cities, but little is said about land developers, real estate syndicators, or insurance companies (for an introductory overview, however, see Feagin and Parker 1990, pp. 63–99). We are just becoming aware of two apparently significant phenomena. The first is the increasing linkage between the financing of real estate development and the broader capital markets. This has occurred simultaneously with the trend toward the internationalization of capital markets. The second is the emergence of new kinds of organizations to plan and execute development projects: these are larger and more closely tied to non-real estate organizations than were developers in the recent past and are more capable of operating in multiple regions, even countries. These two phenomena constitute what I understand as the globalization of real estate.

I will describe these trends here in some detail. My conclusions from this description will be tentative, because we just do not know enough about what is happening. But I will argue that globalization progresses in fits and starts and that what we experienced in the early 1980s has already been partly reversed. I take the position that the changes referred to as globalization do not diminish the importance of local decision-making about land development, however much they change the environment in which local officials operate. In short, I think that reports of the demise of localities are much exaggerated.

Changes in financing

Real estate development is a highly leveraged industry. This means that development firms have traditionally operated with little of their own capital, borrowing instead against the value of their projects at completion. Typically this financing is initially offered by lending institutions as a short-term bridge loan for site acquisition and construction. On completion, or when the property is sold, longer-term financing is arranged.

There have been radical changes in financial mechanisms in the past decade. Historically, savings and loan associations (S&Ls) provided about 50 percent of residential mortgage funds, but already by late 1981 their share of the mortgage market had dropped to 8 percent (ULI, 1983, p. 35). This reflects the damage done to the savings and loan association industry during the 1970s, when it was tied to long-term low-rate mortgages during a period when they had to pay high interest rates to attract depositors. Replacing the S&Ls have been pension funds, life insurance companies, and large commercial banks. At the same time, S&Ls moved precipitously into non-residential projects, hoping to make quick profits to shore up their capital reserves, with disastrous results.

Equally important has been the development of new mechanisms to link real estate financing more directly to the broader capital markets, through a process that is called 'securitization.' Historically, many kinds of investors have been willing to invest in stocks and bonds. These 'securities' had the virtue of being easy to trade (in the stock exchange). Their prices were well known, established by millions of transactions every day, and it was relatively easy to evaluate the strength of the companies that stood behind them. Real estate, on the other hand, has historically been a risky investment. If you invested in an office building, for example, you were tied to the fate of that particular building, in a particular place, with little information about where the 'market' was headed. This is one of the reasons why local investors tended to do better in real estate—they had connections and knowledge of the local scene.

Securitization of Real Estate

For housing developments, semi-public corporations have been established to link the mortgage and capital markets. Single-family homes are not difficult to appraise (since there are typically many sales of comparable homes in a large neighborhood every year). Furthermore, banks have always used the financial condition of the home buyer as a criterion for granting a loan. A creditworthy buyer and an efficiently appraised home combined to make a home mortgage relatively safe. Like stocks and bonds, mortgages could be easily traded through semi-public agencies like FNMA and FHLMC, the Federal National Mortgage Association and the Federal Home Loan Mortgage Corporation. These agencies buy home mortgages in bulk from the institutions that 'originate' them. They, in turn, sell bonds based on the value of the mortgages, attracting capital from all sorts of investors. This process is referred to as securitization. The home becomes collateral for a loan that can be publicly traded at an attractive interest rate.

This process has affected the way that residential construction is financed. Larger home builders have found that they could bypass local lenders completely by arranging financing packages via mortgage bonds directly with FNMA. Securitization not only attracted new sources of capital into the housing market, but also gave new advantages to large builders.

Private sector markets have been established to achieve the same kind of linkage to the broader capital market for commercial and industrial developments. In the 1980s, a number of new mechanisms were established to introduce real estate assets into the securities markets. These involve securitization—converting an asset into a financial obligation that has readily identified characteristics and can be accordingly rated to risk in the international capital markets. This can be done through rating services (the same Standard and Poor's that rates corporate bonds) or credit enhancement by a third party (a form of insurance, for a fee). Japanese banks have been very competitive in this process due to their own high credit ratings, low funding costs, and the low fees they have been willing to charge in order to promote their global business expansion.

Part of the trick is to take advantage of knowledge of financial markets around the globe, and the differentiated preferences for risk, term, and currency (partly due to varying judgements about the future) of alternative investors. If this is done successfully, short-term financing at the lowest possible financing cost anywhere in the world can be replaced by long-term fixed rate financing in dollars, appropriate for a U.S. real estate project. The other part of the trick, of course, is to convince investors that a real estate security is just like any other financial instrument.

Insurance and pension fund investments

This was facilitated in the last decade because of the large demand for investment opportunities in real estate, especially through insurance companies and pension funds. With the disappearance of investment opportunities in oil, gas, and agriculture, and having suffered major losses in loans to Third World countries, lenders committed to the real estate community 'to keep the asset side of the balance sheet from withering' (Zell, 1986, p. 4). Public pension systems pressured for greater authority to make such investments. For example, in 1986 the Illinois legislature revised regulations to allow non-Chicago public funds to invest up to 10 percent in real estate and some other previously restricted categories (Covaleski, 1987). The $1.3 billion Baltimore Employees' Retirement System had 4.5 percent of its assets in real estate, with a target of 10 percent (Hemmerick, 1987). Coincidentally, this was the same period in which 'junk bonds' were becoming popular among institutional investors. On paper, a junk bond offers a high interest rate; the risk level is so high, however, that the rating agencies do not consider them 'investment quality.' Nevertheless billions of dollars were pulled into the junk bond market.

Prudential Life Insurance, itself a major investor in real estate on its own account, promoted real estate investments through its PRISA fund, which amounted to $3.7 billion in early 1989 (Hemmerick, 1989). For example, in 1987 the $7.6 billion Western Conference of Teamsters Pension Trust (Seattle) had all $800 million of its real estate allocation in various Prudential funds (Hemmerick, 1987).

New devices were created to attract pension investments into real estate by assuring security. One, first offered in 1986 by Equitable Life Assurance Society and Merrill Lynch, Hubbard Inc., was a guaranteed real estate investment contract (GREC) that invested in real estate and provided a guaranteed minimum return of at least 8.75 percent (Covaleski, 1987). Another new device was mortgage-backed securities based on jumbo mortgages. Jumbos are loans exceeding the maximum allowable amounts by the Federal Home Loan Mortgage Corporation. In 1986, the jumbo market had reached 22 percent ($96.8 billion) of the total mortgage originations of $440 billion. These were being packaged by a number of banks and mortgage companies for sale to institutional investors.

Direct investments by pension funds

Through such mechanisms, insurance companies and other financial intermediaries handled the bulk of pension fund investment in real estate in the early part of the decade. By the mid-1980s, however, there were large increases in the pension funds directly committed through equity real estate managers: from $6.9 billion in 1986 up to $10.9 billion in 1987. One cause was the increase in stock market prices prior to the 1987 market crash; this forced growth in real estate investment becuase asset managers were committed to target allocation formulas. Another cause was dissatisfaction with performance by 'commingled' investments (such as PRISA). The $1 billion Orange County Employees Retirement System (with an allocation goal of 15 percent to real estate) was planning to withdraw partially from PRISA in 1987, in favor of hiring its own separate account managers. This created a new kind of large-scale entrepreneur able to operate on a very large scale using other organizations' money: JMB Realty ($1.9 billion in assets), Copley Real Estate Advisors ($1.2 billion), Heitman Advisory ($1.1 billion), Equitable Real Estate Investment ($1.0 billion). In a single deal in 1987, two of these firms together raised $2 billion from pension funds to finance the purchase of Cadillac Fairview of Toronto—one of the world's largest development companies—by Canadian developers Olympia and York (Hemmerick, 1987).

Capital market activities of development companies

Another change in the 1980s was the ability of large developers to raise capital directly without the intermediary of an insurance company, investment fund, or asset manager. An early case occurred in November 1985, when Lincoln Property Associates Ltd. (the nation's second largest real estate developer at that time) raised $146 million through the sale of 15-year participating first-mortgage bonds on commercial office building projects. This deal was underwritten by Drexel Burnham Lambert Inc., which was a pioneer at the same time in junk bonds for leveraged buyouts. It was estimated that the total return would be about 14 percent. This allowed Lincoln to expand its investor base beyond a few major institutions; only five large institutions, including Metropolitan Life Insurance Company and TIAA, had been buyers of such bonds through the private market (Baker, 1985).

Even before this, in 1983, Olympia and York (whose operations are examined in detail in Chapter 6) had pioneered the use of mortgage-backed securities for commercial properties. With Salomon Brothers, Inc., they created a $970 million blanket mortgage on three office buildings and offered securities to the public based on this mortgage in 1984. The purpose was to obtain better rates than available through a private placement and limit potential liabilities by making it a 'non-recourse' loan—only the mortgaged building could be taken over in case of default (Scardino, 1987).

Such deals were greatly facilitated by the announcement by Standard & Poor's in late 1984 of criteria for rating securities based on individual commercial properties. Medium-sized investors, without special real estate expertise, were satisfied by such ratings. In April 1986, Olympia and York entered the Eurobond market with a $485 million public offering backed by a mortgage on 55 Water Street (Manhattan). This was 'the largest rated nonrecourse real estate financing ever' (Meyers, 1987). It received a double-A rating from Standard & Poor's, and the bonds were successfully marketed by Salomon Brothers. The world's investors put up nearly half a billion dollars in return for pieces of paper whose value depended entirely on the future of the Manhattan real estate market.

Changes in organizational relationships

At the same time as the transformation of real estate finance, and partly resulting from it, there have been several changes in the network of organizations involved in development. These include an increasing scale of development firms, as well as various forms of horizontal integration in which development firms gained more direct and continuing access to the capital market.

The scale of development organizations

It is natural that the increasing scale of available financing resulted in the appearance of development firms of unprecedented capacity. Several are mentioned specifically here, because they can be used to illustrate key aspects of the operation of such firms.

The largest privately owned firm, Olympia and York, is based in Canada but has substantial operations in the United States and Great Britain. In 1986, Olympia and York Developments owned more than fifty million square feet of space in North America, including the World Financial Center in Manhattan (Foster, 1986). At that time they were planning their largest project outside North America, the $6.5 billion Canary Wharf redevelopment project in London, which is currently under construction and will create more than 12 million square feet of office space (see Chapter 6). Olympia and York began diversifying out of real estate in the early 1980s. For example, in 1985, they bought 60 percent of Gulf Canada from Chevron for $2.8 billion. In 1986, through Gulf Canada, they paid $3 billion for a 69 percent interest in Hiram Walker Resources. Clearly the distinction between real estate and non-real estate ventures is blurred in this case. But other major acqisitions involved other large-scale real estate developers: Cadillac Fairview, Landmark Land, Campeau Corporation and Trizec. And Olympia and York's investments in Santa Fe–Southern Pacific were at least partly motivated by the railroad's substantial land holdings throughout the United States.

The largest developer in the United States for most of the last decade has been the Trammel Crow organization. Trammel Crow was established in the early 1950s as a warehouse developer and quickly dominated that special sector. From near bankruptcy in 1976, Crow grew from assets of approximately one billion dollars to three billion in 1982 and 13 billion in 1986. From 15 offices in 1976, there were 90 in 1986, while the number of employees grew from 300 to 5,000 (Sobel, 1989, p. 210).

Trammel Crow is so large that even its offshoots became nationally prominent. For example, Lincoln Land Company was created in 1965 as a spinoff from Trammel Crow to build garden apartments. By 1986, it was building at the rate of about $1.6 billion per year, having sold off some assets but retaining $4.7 billion of properties, with estimated equity of $1.4 billion (O'Reilly, 1986).

Horizontal Integration

Three types of changes in organizational relationships merit special attention: the establishment of financial subsidiaries by development firms, the creation of long-term financial partnerships between developers and insurance companies, and the capture of savings and loan associations by real estate developers.

1. *Creation of subsidiaries*
A good example of the creation of subsidiaries of development firms is provided by Trammel Crow. As early as 1959, Crow created a new company, Wallace Properties, Inc., to assist in development financing. In 1961, Wallace acquired Gibraltar Investment Company of Los Angeles through a stock swap. The primary asset of Gibraltar was Institutional Mortgage Company, which invested in Federal Housing Administration and Veteran's Administration mortgages. In 1961 the company changed its name to Wallace Investments and established Wallace Realty Mortgage as one of it subsidiaries. In 1963, Realty Mortgage acquired Lomas and Nettleton, a Connecticut-based mortgage banker, as well as Admiral Fire Insurance of Houston (Sobel, 1989).

2. *Direct equity investments by financial corporations*
Sometimes lenders demand a share of equity in a project in return for lower interest rates, giving some of the financial benefits of equity to the debt holders. So there is a blurring of debt and equity. Sometimes financial institutions such as life insurance companies, who once offered the main source of permanent financing for commercial developers, now compete directly in the development market.

Lincoln Land Company provides an example of direct investment through partnerships by a major insurance company (O'Reilly, 1986). Metropolitan Life invested $2 billion, about 10 percent of its total real estate investments, through Lincoln during 1979–86. In a typical deal with Met Life, Lincoln would spend about $2 million in startup costs and be liable for construction cost overruns. Met Life would pay off the short-term bank loan covering land acquisition and construction costs, plus a 2–3 percent 'development fee' to Lincoln, in return for 50 percent ownership. Lincoln would make fixed payments for half the construction costs, plus interest, over 15 years; if rental income were sufficient, payments on the other half of the loan would be higher. The remainder of rental income and capital gains when the building was sold would be shared by Landmark and Met Life. But if income were insufficient, Lincoln would have no liability beyond forfeiting equity in the building to Met.

Such an arrangement is clearly advantageous for a development firm, which receives priority access to capital without liability. For the insurance company, it is advantageous only if the long-term cash flow and capital gain are positive. Met Life stopped investing in high-rise office construction in 1985, concerned about high vacancy rates around the country. But in its place, Lincoln found new financing through mortgage-backed bonds issued by Drexel Burnham Lambert (again, famous for junk bonds). In a $146 million bond issue in December 1985, investors got 15-year bonds with a first mortgage on the properties at 10.5 percent interest, plus a share of rental income and capital appreciation.

Bonds were used as vehicle to replace a single large insurance investor with many smaller investors, still on the premise of shared equity.

Prudential Life Insurance operated on an even larger scale, with more direct control of its real estate investments through the Prudential Realty Group (PRG). Established in the 1960s, PRG was by 1987 clearly one of the world's largest private real estate investors. Ownership assets handled through PRG included nearly 100 million square feet of office space, 80 million square feet of industrial, 25 million sqare feet of retail, 37,000 hotel rooms, 11,000 apartments and condominiums and 800,000 acres of farmland. The value of assets administered by PRG exceeded $34 billion, nearly equally divided between mortgages and real estate equities (Karsian, 1987, p. 98).

3. *Capture of savings and loan institutions*

Another far-reaching change in organizational relationships was stimulated by federal and state legislation to deregulate savings and loan associations in the 1980s. The Garne–St. Germain Depository Institutions Act of 1982 made savings and loans attractive acquisitions by expanding their asset and liability power. Some states went further: state-chartered S&Ls in California were permitted after 1982 to create a wholly owned service corporation holding 100 percent of its assets, with the power to conduct virtually any business activity, including real estate development, syndication and contract construction. California's Nolan Bill of 1982 also made it easier to start a new S&L and provided for extremely profitable financial leveraging. Around 200 applicants filed for California charters in 1982 and 1983, many of whom were individuals who operated land development, home building and real estate syndication companies.

Deregulation was put forward as an effort to allow savings and loans to find their own way out of financial difficulties by making loans with greater returns than traditional residential mortgages. In retrospect it is obvious that the risks of this strategy outweighed the returns, with long-term costs to government agencies of up to $500 billion. What is interesting here is the way that deregulation created new ties between developers and S&Ls. Two examples are given here: the Vernon S&L case that has already resulted in successful prosecution, and the activities of Landmark Land, mentioned above in relation to Olympia and York.

In 1982, Don Dixon (a Dallas condominium builder) acquired a 90 percent interest in Vernon Savings and Loan for $5.8 million, mostly borrowed. Seeking funds through national deposit brokers who delivered huge sums in return for higher than standard interest rates, Dixon increased Vernon's deposits from $82 million in 1982 to about $500 million in 1983 and $1.7 billion in 1987. He focused on investments in construction loans and other deals with no down payments, so that by 1985, 70 percent of Vernon's interest income was from land purchase, development and construction loans. Loan fees were twice the industry standard: 4 percent for origination, and 1–2 percent for renewal every 6 months; risks were commensurately high.

Vernon's activities illustrate several sorts of gimmicks that were used in the 1980s to maintain streams of financing to high-risk projects:

- 'Cash for trash' provided borrowers with cash for a requested loan in return for agreement to take out a second loan to buy troubled real estate properties owned by the lender, covering up bad loans on those properties. In 1984 Dixon learned that real estate owned by a Vernon subsidiary was worth $20 million less than book value. He concealed this by lending funds to associates to buy these parcels, and showed a $13 million profit instead of a $10 million loss (O'Shea, 1988b)!

- Another gimmick was to transfer loans to the books of other institutions through 'back scratching.' In the mid-1980s, Vernon was growing faster than allowed (taking on liabilities too much in excess of its own capital). It avoided scrutiny by bank examiners by selling 'loan participations' valued at more than $450 million to other S&Ls. When Vernon was closed in 1986, with estimated losses of its own of more than $600 million, these loans became losses for the other firms.

Other practices have been documented in investigations of fraud in the thrift industry, most of which had the consequence of allowing developers to take advantage of captive institutions. 'Back scratching' was a variant of another method of concealing losses. A savings and loan would make a loan at 120 percent of what was needed and place the extra 20 percent into an 'interest reserve' account to pay the interest for the first two years. This would show the loan to be profitable for two years. Then it would be refinanced by another S&L. To gain loans for projects in excess of their real value, developers engaged in 'land flips'—consecutive sales of land at inflated prices to establish an appraisal for loans. In one Dallas case, it was discovered that FHA's own appraisers were implicated in such transactions.

Fraud was not, nevertheless, an essential ingredient in the manipulation of S&Ls by developers. An example where bank regulations themselves authorized the maneuvering is Landmark Land, one of the country's largest publicly owned land developers in 1987, with a value of about $180 million (25 percent owned by Olympia and York, 30 percent by Gerald Barton). Barton took over Landmark in 1971 and developed several housing subdivisions and golf club communities in the Sunbelt. In 1982, he purchased Dixie Savings and Loan, the largest in Louisiana, from the Federal Savings and Loan Insurance Corporation. Dixie had failed as a result of failure of investments in energy-related loans and a depressed real estate market.

With agreement by FSLIC, Barton recapitalized Dixie without injecting new cash reserves, simply by putting all of Landmark's land on Dixie's books. This increased Dixie's authorized borrowing capacity. Barton then converted from a federal to a state charter that allowed him to invest up to 10 percent of the expanded asset base in real estate. Accounting rules prevented evaluating the land assets at current market value, so that Dixie was limited to real estate investments of $222 million. To circumvent this, Barton bought another failed Louisiana thrift, St. Bernard Savings and Loan, in 1986, and transferred all of Dixie's real estate assets to St. Bernard at current value. This increase in capital enabled St. Bernard to make real estate investments up to $1.5 billion, representing a valuable financial tool for Landmark Land (Paris, 1987).

Cycles or trends?

Let us recapitulate the several processes that have been referred to thus far as the 'globalization' of real estate. On the financial side, securitization is the critical innovation that has allowed local property development to be financed in the national and international capital markets. Closely related to securitization is the expanding range of potential investors, particularly pension funds, and the growing role of intermediary financial institutions such as insurance companies and realty asset managers to channel these investments. There has also been a trend toward direct entry into the capital markets by large development firms, with the assistance of investment bankers.

There have been corresponding changes in the interorganizational network involved in property development. First, the market has become more concentrated, with the emergence of a number of mega-developers—some operating in the full range of markets and others specialized in housing, warehouses, shopping centers or office buildings. These firms are positioned to take advantage of the demand for real estate investment opportunities and to operate at a national and international scale. Three types of changes in their relationships with financial institutions have been described: the establishment of financial subsidiaries by development firms, the creation of long-term financial partnerships between developers and insurance companies in which the latter have an equity position and the capture of savings and loan associations by real estate developers. The globalization of real estate has been accompanied by various forms of horizontal integration, blurring the traditional distinctions between developers and financiers.

What are we to make of these changes? Are they portents of continued restructuring of the real estate industry, its multinationalization in the same sense that manufacturing has become multinational? Some participants in the industry itself have argued that the changes are of limited scope. First, although development firms operate in national and international markets for capital, most property development is conducted–indeed, much is originated–by local partners. Second, securitization does tend to segment property development into national products, which investors believe can be rated to risk, and local products, which cannot. The rating process is itself nevertheless fraught with risk, and further extension of securitization should not be taken for granted.

The role of partners

In actual operation, most major multilocational development firms operate through local or regional partners. These partners offer expertise on specific market conditions (including land prices and availability, and potential buyers or renters), and contacts with the network of construction companies, architects, and suppliers that will be involved in the development. They are often especially important for the interface between the national developer and local government.

Consider the situation of North American investors in Europe, for example (the following quotes are from Hylton, 1989). Charles Grossman, managing director of an international consulting firm, predicts that 'only the leading American developers are going to make it in Europe. It's hard to find sites and hard to get consent to build. You're not going to go to London and wave your magic wand and build a million-square foot building.' Debra McClain, head of Morgan Stanley's European real estate operations agrees: 'The big developers are working in some way with local expertise, either with local consultants or with local developers as partners . . . You need to know who your competitors are, and you need to know the local authorities to get through the planning process.'

I provide two examples here of multi-billion-dollar developers who make extensive use of local partners in the United States: Trammel Crow (one of the largest nationally) and Pyramid (which operates only in the Northeast).

1. *The Trammel Crow organizaton*

Trammel Crow operated from the beginning through decentralized local partnerships. As his operations grew in number, projects were pursued in connection with different sets of partners, generally through separate and unrelated agreements rather than a centralized corporate structure. The typical initial arrangement was an 80-20 split, rising to 50-50 or better as the new partners took on additional responsibilities. A short-lived partnership was with John Eulich, which ended in 1963 (after which Eulich organized Vantage Corporation, which became one of the nation's largest real estate developers). Eulich's replacement in the residential business was Mack Pogue, a Dallas real estate broker, with whom Crow created Lincoln Property (discussed above), which built apartments around the country through 1976. Expansion beyond Dallas was accomplished by relocating three of Lincoln's Dallas partners in Houston, St. Louis and San Franciso, making each person a regional partner, with satellite companies under his own control.

Despite its advantages, operating through a complex set of partnerships eventually caused its own problems. In 1976, the lack of accountability between partnerships and Crow's funnelling of cash flow among them resulted in a close brush with bankruptcy. At that time, the Crow empire began to move toward a more centralized and routinized organizational structure. The issues of partner sovereignty and relative power of central and regional forces and conflicts between Crow's family holdings and those involving regional partners all combined to constitute an extremely tense network. A large number of local partners have subsequently broken away to establish separate organizations of their own, despite the apparent advantages of scale in Trammel Crow.

2. *The Pyramid organization*

Pyramid Corporation is a large-scale developer of shopping malls in the Northeast with assets of about $2 billion. According to Gratz (1989), it is fully controlled by the Robert Congel family. 'Every Pyramid project is a separate general partnership in which Congel or a member of his family have a controlling interest, which varies from 51 to 75 percent. There are no limited or outside partners but each project usually has six to nine general partners . . . and one sponsoring partner, who is the site manager, the one who ushers the mall proposal through the public approval process, who hires influential local lawyers, architects, engineers and environmental consultants' (Gratz, 1989, p. 54).

Pyramid illustrates the political motivatons for local partnerships, and the selection of both local business people and well-connected politicians to represent them locally. One important case was in the small city of Poughkeepsie, New York, where Pyramid's 1.1-million-square-foot Galleria opened in August 1987. The Democratic majority on the Town Board had opposed a needed zoning change for this mall. Consequently, in the town elections of 1985, Pyramid funnelled over $750,000 to pro-mall Republican candidates–'neither the candidates nor the public knew where the support money was coming from' (Gratz, 1989, p. 58). Comparable elections had typically involved total contributions of under $20,000. Apparently these contributions were legal, and Congel defends his firm's local intervention: 'There is always a certain amount of business that takes place under the table. There's hardly an election where cash does not change hands. We knew ahead of time that either we left Poughkeepsie or did it legally and played by the rules .. . . It's a choice we made, and I'm proud of it' (p. 59).

Another case is in Williston (near Burlington, Vermont), where Pyramid has proposed a 400,000-square-foot mall on 72 acres, near which Pyramid owns a 160-acre farm. Their original proposal was defeated in 1978 due to strong local environmental opposition. Now Pyramid has recruited a local businessman, Ben Frank, as a sponsoring partner. An opponent (Chico Lager, an executive of Ben and Jerry's Ice Cream) describes the strategy this way: Pyramid consultants who evaluated their Poughkeepsie experience concluded that they 'didn't need to go through such gyrations and spend so much money to buy an election. They could play it smarter by being Mr. Nice Guy with good public relations and a polite front man (a la Ben Frank) to talk straight to the people . . . While they always had a local partner carrying the ball for Pyramid, now they have refined the role and look for a representative who will appease rather than wear down and challenge a local community' (Gratz, 1989, p. 59).

In brief, these cases show that in some respects land development continues to have a substantial local character even when carried out by nonlocal firms. Certainly the entire development process now includes financial and political dealings at the state and national levels. However, the local dimension is not bypassed but incorporated into this larger network.

Harvey Molotch and I have discussed elsewhere the special resources available to large-scale developers as they maneuver at the local level (Logan and Molotch, 1987, p. 228–44). These include the ability to assemble projects in which profits from externalities are reaped by the firm itself, the support of local managers and growth machine partners and the potential to bring substantial political pressure to bear from higher levels of government. None of these, it seems to me, appreciably undermines the necessity of being able to operate locally. The parochial participants in city growth machines may tend to become the junior partners of extralocal organizations, but they continue to have a critical role. In pursuit of their own more modest plans, with local backing, these may in fact be the principal players in some places and at some times.

Market segmentation: national and local markets

What are the relative roles, and prospects, of national and local developers? Clearly it is difficult for small-scale builders to compete successfully with the big developers, who go directly to the capital markets to obtain financing. Following this logic, the Urban Land Institute argued in 1983 that 'successful developers during the remainder of this decade will not only have a marketable product, but will also be strongly connected to reliable sources of development capital. As a result, entry into the development industry will necessarily be more restricted' (1983, p. 7).

One specific advantage of large developers is their ability to handle projects whose securities can be sold on a national market. These include 'projects large enough to capture a niche in the local market, new enough to have a competitive advantage, and marketable enough to meet expectations of growing or stable demand' (Sears, 1983, p. 315). The real estate equities market is based on prices that are difficult to establish, requiring detailed title searches and insurance, with complex tax implications. Smaller investors, therefore, find their competitive advantages in more local development projects and remain a source of local capital for such projects.

Another potential advantage of large firms is their ability to withstand cyclical fluctuations in land values, rents and the cost of funds. If financial institutions adopt a conservative posture toward real estate, avoiding long-term and fixed-rate loans, more developers or owners will be forced to sell properties, especially in adverse times in the interest rate cycle, when prices are depressed. For those without 'deep pockets' real estate will become more risky. A great deal of working capital is required to overcome price fluctuations and variable marketability of real estate at different points in the business cycle (Sears, 1983).

Yet one market insider, Samuel Zell (1986, p. 1) concludes that the 'commoditization' of real estate and its exchange on national and international markets will eventually result in 'one of the biggest losses of capital in the country's history.' He argues that real estate is a non-fungible asset: one property cannot be simply exchanged for another as though both were the same product (this is part of what Logan and Molotch [1987, p. 23] refer to as the 'specialness of property as a commodity'). 'Real estate is a local market, by definition. It is not possible to focus on national trends; one must focus on local issues and characteristics' (Zell, 1986, p. 1).

Securitization, Zell argues, only blurs the risk taken by the investor; it does not erase it. Further 'abstract fund allocation [the targets described above for institutional investment in real estate] continues the thesis of distancing the real estate participant from the property. . . . The real estate business . . . is a highly leveraged business that requires an attention to detail that does not lend itself to delegation. The conversion of real estate from a localized to a national business has . . . led to the greatest oversupply of brick and mortar in the country's history.' (Zell, 1986, p. 4).

Five years later, there is considerable evidence to support Zell's view. Many of the large development firms that emerged in the 1970s and early 1980s, partly in response to capital availability, have experienced severe financial setbacks in the late 1980s. And the deregulation of the banking industry, which allowed S&Ls to become more closely coupled with speculative real estate investment, has instead promoted the near collapse of this industry.

The failure of the large development firms in the late 1980s

Already by 1985 there was ample evidence of overbuilding in office space, which led such institutional investors as Prudential to halt new funding of such projects. In the highly concentrated housing industry, four of the ten largest housing builders in 1986 (in terms of number of starts) had recently completed restructuring or were in the midst of it; one was in Chapter 11 bankruptcy proceedings (*Wall Street Journal*, 1987). These included No. 1 Trammel Crow of Dallas (14,545 starts), whose reorganization was provoking the resignations of several regional partners; No. 4 U.S. Home of Houston (8,494 starts), which had reduced its scale of operations due to poor performance in the Southwest; No. 7 Lincoln Property of Dallas (7,266 starts), which was attempting to resposition its market from the Southwest to the Northeast; and No. 10 Nash Phillips/Copus of Austin (6,157 starts), which went into Chapter 11 protection in March 1987. Massive size, in the housing sector, was clearly not protection against massive losses.

For these large developers, in one analyst's view, the most promising future would be as 'merchant builders for the insurance companies or pension funds . . . Because real estate seems to be getting back to what it really always has been: a local business' (Taylor, 1989, p. 100).

The failure of the savings and loan industry

Concurrent with the problems of development 'dinosaurs' (as Taylor called them), the late 1980s also witnessed a record rate of failures and loan associations whose loan policies had fueled speculative development. By the end of 1987, insolvency losses to FSLIC had reached $30.24 billion in Texas alone, followed by California ($9.38 billion), Illinois ($2.31 billion), Flordia ($2.25 billion), New York ($1.61 billion), Louisiana ($1.58 billion), Oklahoma ($1.51 billion) and Arizona ($1.24 billion). The number of failed institutions was 70–80 per year in 1986–87, up from about 10 in 1980. In 1988, more than 500 were being kept open by FSLIC in order to avoid paying off the losses (O'Shea, 1988a).

The federal government, through FSLIC, attempted to postpone a final reckoning through temporary arrangements. Under a plan announced in February 1988, the number of S&Ls in Texas would be reduced from 281 to 160–180. 'Under the plan, insolvent institutions would be combined with solvent ones. The surviving institutions would be bolstered with subsidies and notes to lure investors and fresh capital, which would enable them to eventually stand on their own' (O'Shea, 1988c). One of the first of these deals was a consolidation of Southwest Savings Association in Dallas (owned by trusts for the benefit of Caroline Rose Hunt) with four insolvent firms. Southwest, with its own assets of $1.48 billion, would take over four institutions (including Lamar Savings Association) with assets of $4.3 billion. FSLIC would own 50 percent of Southwest's common stock and get 90 percent of the first $60 million in profits (though currently Southwest was losing money). In return, Southwest–'which has negligible capital'–received a $438 million note and other subsidies with a value of at least $1.5 billion.

O'Shea notes that no new private capital had been brought in, and the interest rates paid by Southwest had not declined. Referring to a total of seven deals under this plan, he cites a savings industry analyst: 'The lack of capital infusion in the transactions means that these are not real or final deals. All the FSLIC has done is move $9.8 billion of negative net worth out of the thrift industry and onto its balance sheet' (O'Shea, 1988c).

Acknowledging after the 1988 Presidential election that such temporary measures would be inadequate, Congress established the Resolution Trust Corporation, with current authority to issue up to $73 billion in bonds to finance mergers and closings of S&Ls. Total cost, including interest on bonds, is now estimated at about $500 billion. More significantly, regulators have sharply curtailed the lending authority of the remaining savings and loans. As of December 7, 1989, they must meet three requirements regarding capitalization: (1) tangible capital (real assets less liabilities) of 1.5 percent of assets, (2) core capital (common equity, retained earnings) of 3 percent of assets, and (3) risk-based capital (core capital, cumulative preferred stock, plus some debt) of 6.4 percent of assets. Of the nation's 2903 thrifts, 648 have been unable to meet these requirements (Odato, 1990). The special relationship between developers and the banking system may have reached its limits.

Conclusion: Global developers in local markets

We should be cautious about the conclusions drawn from the globalization of real estate. Some of the trends of the past decade depended upon the accumulation of capital in the hands of insurance companies and pension funds, the closing of several alternative directions for investment, optimism in future gains in land prices and rising rents, deregulation of financial institutions and particularly the availability of savings and loan associations as adjuncts of development firms. Failures of some large developers in their current form and the crisis in savings institutions are sufficient to cast doubt on the continuation of some of these conditions. The recession which began in the 1990 will further erode the system that evolved in the 1980s, although political pressure is currently being applied to 'relax' regulatory standards.

There have been changes in financing and in the network of organizations involved in development which surely will have long-term effects. The more closely real estate finance is tied to the general capital markets (this trend has been sustained), the more the cost of real estate funds will reflect general variations in interest rates. Therefore we may anticipate greater volatility in rates and accentuation of the usual cycles of expansion and contraction in real estate development. Also, because funds on the open market (as compared to local investments) concentrate in areas of perceived low risk and high return (as was the case in the Sunbelt in the late 1970s), globalization will promote overinvestment in some markets and reinforce regional cycles of growth and decline.

Both interest rate volatility and regional imbalances will contribute to an aggravation of what Harvey has described as 'switching crises' in capitalism (Harvey, 1978). Their consequences will be felt most strongly by the least protected classes of people and places, for whom funds are typically only marginally available.

Depending on the ability of the global market to assess risk and continue to securitize real estate investment, we can anticipate that the distinction between a national and local tier of projects will remain. This will tend to draw funds toward larger projects and away from those conducted on a smaller scale. And related to this segmentation, the trend toward direct equity investments by major institutions such as life insurance companies and pension funds will reinforce the capital market advantage of larger projects which constitute a more manageable portfolio.

For all of these reasons, publics and policymakers at the local level will increasingly be confronted with mega-developments with the potential of making substantial changes in the built environment. With the apparent power to direct investments elsewhere, global developers have strong bargaining leverage with localities. But, as noted above, there are good reasons for them to operate through partners who are based in the locality and to continue to have to deal with local officials. As shown here, in some respects both their ability to invest 'anywhere' and their 'global' organization are more apparent than real.

References

Baker, M. 1985 'Real estate company aims to develop public market,' *Pensions and Investment Age,* November 11, p. 69, 74.

Covaleski, J. 1987 'Guaranteed contract draws 25,' *Pensions and Investment Age,* August 24, pp. 15, 17.

Domhoff, G. W. 1983 *Who Rules America Now?* Englewood Cliffs, N.J.: Prentice Hall.

Feagin, J. and R. Parker. 1990 *Building American cities: The urban real estate game.* Englewood Cliffs, N.J.: Prentice Hall.

Foster, P. 1986 *The Master Builders: How the Reichmanns Reached for an Empire.* Toronto: Ekey Porter Book Limited.

Gratz, R. B. 1989 'Malling the Northeast,' *New York Times Magazine.* April 1, pp. 35–59.

Harvey, D. 1978 'The urban process under capitalism: A framework for analysis,' *International Journal of Urban and Regional Research*, 2, 101–31.

Hemmerick, S. 1987 'Investors move away from broadly diversified funds,' *Pensions and Investment Age,* August 24, pp. 15, 17.

Hemmerick, S. 1989 'PRISA rebounds from record lows,' *Pensions and Investment Age,* March 6, p. 24.

Hylton, R. D. 1989 'Developers rushing into Europe,' *New York Times*, October 10, pp. D1, 4.

Karsian, D. 1987 'Ethics, professionalism top priority at Prudential; realty reorganizaion to focus on overseas markets' *National Real Estate Investor*, 29 (February), pp. 98–104.

Logan, J. R. and H. L. Molotch, 1987 *Urban Fortunes: The Political Economy of Place.* Berkeley and Los Angeles: University of Caliornia Press.

Logan, J. R. and T. Swanstrom, 1990 *Beyond the city limits: urban policy and economic restructuring in comparative perspective.* Philadelphia: Temple University Press.

Meyers, W. 1987 'Wall Street takes the plunge' *Institutional Investors*, 17 (January), pp. 135–140.

Molotch, H. L. 1976 'The city as a growth machine: the political economy of growth,' *American Journal of Sociology*, 84, 309–32.

Odato, James. 1990 'Holding Co. Northeast Savings goal,' *Schenectady Gazette*, June 30, p. C13.

O'Reilly, B. 1986 'This builder wants it all—without risk' *Fortune*, May 12, pp. 50–57.

O'Shea, J. 1988a 'Taxpayers target of financial 'time bomb',' *Chicago Tribune,* September 25, pp. 1, 28.

O'Shea, J. 1988b ' 'New entrepreneurs' build massive S&L losses,' *Chicago Tribune*, September 26, pp. 1, 12.

O'Shea, J. 1988c 'Keeping afloat: taxpayers sure to lose 'shell game',' *Chicago Tribune*, October 2, Section 4, pp. 1, 4.

Paris, E. 1987 'Just so much popcorn,' *Forbes,* 139 (June 1): pp. 44, 48.

Scardino, A. 1987 'Building a Manhattan empire,' *New York Times,* April 18. pp. 27, 29.

Sears, C. E. 1983 'Trends in Real Estate Finance,' pp. 305–328 in Urban Land Institute, *Development Review and Outlook 1983–1984*. Washington, D.C.: Urban Land Institute.

Sobel, R. 1989 *Trammel Crow: Master Builder. The Story of America's Largest Real Estate Empire.* New York: John Wiley and Sons.

Taylor, J. H. 1989 'The dinosaurs are dying,' *Forbes*, 143 (May 1), pp.92–100.

Urban Land Institute. 1983 *Development Review and Outlook 1983–1984*. Washington, D.C.: Urban Land Institute.

Wall Street Journal. 1987 'Trammel Crow No. 1 among builders,' *Wall Street Journal* (April 22), p. 35.

Zell, S. 1986 'Modern sardine management,' *Real Estate Issues*, 11 (Spring/Summer), pp. 1–5.

Three

THE TURBULENCE OF HOUSING MARKETS: INVESTMENT, DISINVESTMENT AND REINVESTMENT IN PHILADELPHIA, 1963–1986

Robert A. Beauregard

University of Pittsburgh

Although the built environment contains many of society's longest-term fixed investments, the commercial areas, industrial districts and neighborhoods of large cities are nonetheless constantly in flux. In a matter of months, façades are repaired or replaced, new stores take up the leases of retail businesses that have failed or relocated, apartment buildings appear on parking lots, buildings are demolished, houses renovated and streets resurfaced. Almost everywhere one finds signs of new construction, abandonment, renewal, or disrepair. Cities are constantly adapting to and being transformed by novel economic activities, the migration of people, large-scale public works, investor speculation, new patterns of everyday life and the desire for change. The landscape is restless because society is inherently unstable. Even the most fixed of capital investments reflect this.

Despite the surface credibility of these observations, our theoretical approaches to the built environment often juxtapose growth with decline, renewal with blight, expansion with contraction. One precedes and displaces the other, rather than existing simultaneously in time and space and thus in a more contingent and problematic relation.

The premise of this chapter is that the built environment is ever-changing and characterized by profound and intersecting spatial and temporal uneven development that barely yields to simple representations of sequence or exclusion. The chapter's purpose is to explore uneven development by analyzing the patterns of investment, disinvestment and reinvestment across the landscape of Philadelphia's housing market between 1963 and 1986. Using a data set which contains information on the construction of new housing units, demolitions of housing units, and conversions to housing units, I investigate how these various processes are interwoven temporally and spatially. My goal is to substantiate empirically and elaborate theoretically our understanding of the 'creative destruction' of capitalism as it works its way through the housing market of one of the country's largest cities.

Uneven development and capital investment

Neoclassical interpretations of temporal and spatial uneven development under capitalism cast its existence as temporary (Richardson, 1969, pp. 347-57; Richardson and Turk, 1985, pp. 5-27; Stillwell, 1978; Watkins, 1980). Greater capital investment in one location rather than another represents investor response to the natural advantages of those locations (e.g. a protected harbor) and to market forces. Given production technologies and prevailing patterns of demand for commodities, investors direct capital into certain locations and not others, thus creating disequilibria in growth rates, profit margins, wage levels, employment rates and population shifts.

These disequilibria, however, are temporary. Capital rushes to locations with low wage rates and profitable investment opportunities, thereby drawing in labor and more capital. As this process unfolds, investment opportunities become fewer, wages rise and surpluses of both capital and labor emerge. Moreover, the scale of development generates congestion, pollution and other undesirable externalities. The result is that the costs related to investment increase, profits fall and labor shortages appear, thus pushing up wage rates. Capital then shifts to undeveloped areas where labor is relatively cheap, investment opportunities have not yet been exploited and negative externalities are minimal. Over time and as geographical expansion reaches its limits, the 'system' should move to equilibrium and different locations should converge, particularly in terms of income levels.[1]

Such a perspective portrays investment in the built environment as sequential. Property development begins in locations that have advantages conferred by the prevailing stage of economic development. Capital flows to these areas and away from areas having relative disadvantages. As overbuilding occurs and densities increase, costs rise and profit margins fall and capital is directed to less developed areas, often on the fringes of built-up areas, but also to new locations that have become more advantageous. Capital thus moves sequentially from areas where land and building costs are high to those where they are low.

The return of capital to central cities during the 1980s would thus be viewed from

[1] The convergence thesis of neoclassical economics is based on an assumption of long-run general equilibrium. Cumulative causation theory posits an opposite model, one with deviating amplifying feedback cycles. Locations that gain an advantage continue to build on it and those disadvantaged fall further behind. See Myrdal, 1957. Wilbur R. Thompson (1968) posits an urban size ratchet whereby large cities achieve a level of development which precludes significant decline, thus limiting their contribution to convergence.

this perspective as an anomaly. Instead of capital flowing to undeveloped areas, it was flowing to previously developed (but now deteriorated) areas. Capital had shunned such central cities during the 1960s and 1970s, preferring suburbia; but suburban locations became costly to develop. Urban locations, because of the fall in property values attendant to neglect and disinvestment generally, became less costly. Capital, then, returned, in violation of the neo-classical analysis. Moreover, the pattern was not solely sequential. Decline in the central cities was also a contributor to growth in the suburbs, not only because decline lowered property values but also because it increased the risks and costs of investment while lowering the likely rates of return. Decline and growth are thereby mutually exclusive as well as sequential, and both were temporary.[2]

On the surface, Marxist interpretations of uneven development seem similar (Clark, 1980; Harvey, 1982, pp. 373–445; Harvey 1985a, pp. 32–61; Scott and Storper, 1986, pp. 195–311; Smith, 1984). They speak of capital flowing to locations with high profit potential and abandoning those with low profit potential. However, the claim of eventual spatial equilibrium is rejected, thereby dismissing the belief that uneven development is a temporary phenomenon.

More important to the Marxist argument is that uneven development is not an unintended and bothersome by-product but an essential element of growth (Browett, 1984). Capitalism requires uneven development to function well. When new areas of growth are exploited to the point where profit rates fall, capital flows to previously undeveloped areas. When stagnation sets in, capital migrates once again, sometimes back to areas previously devalued. In fact, the exploitation of growth areas often requires that capital be freed from previous investments. Areas where profit rates are falling experience disinvestment and thus further devaluation. In addition, profitable investment opportunities are not simply consequences of overall growth and changing demand. Rather, investment opportunities are frequently created through the interplay of capitalists and governments. Uneven development is thus socially produced, not a naturally occurring condition. As a result, the Marxist approach, quite contrary to the neoclassical approach, casts disinvestment as essential to new investment and reinvestment, thus paralleling Schumpeter's (1950, pp. 81–86) belief in capitalism's 'creative destruction.' In general, and speaking ideologically, new capital investment is always suspect, disinvestment is viewed as insensitive to social needs, and reinvestment is evidence that earlier disinvestment had been functional and purposive.

The application of the Marxist interpretation of uneven development to the built environment requires that we distinguish between the spatial sequentiality of investment

[2] Bradford and Rubinowitz (1975) discuss these issues, though not from a neoclassical perspective.

and its spatial simultaneity.[3] Capital does switch from one location to another. To this all would agree. More contentious is the explanation as to why this occurs and thus the relationship between investment and disinvestment. A metaphor used by Neil Smith conveys that relationship in very vivid terms: 'Behind the extant pattern of uneven development lies the logic and the drive of capital toward what we shall call the 'see-saw' movement of capital' (Smith, 1984, p. 148). In effect, '. . . capital attempts to see-saw from a developed to an undeveloped area, then at a later point back to the first area which is now underdeveloped, and so forth . . .' (p. 149). The image is one of sequentiality. Investment follows disinvestment and disinvestment follows investment in a temporally linear sequence. The reason is that capitalists need to disinvest in certain areas in order to free up capital for investment in other areas. Thus, for instance, capital abandons existing housing or factories in declining areas in order to have investment capital for new housing or factories in growing areas (Harvey, 1985a, pp. 1–31).

Contrast this with an interpretation, also contained in the writings cited above, that posits a spatial simultaneity. Here the see-saw metaphor is abandoned. Investment and disinvestment can occur simultaneously within the same location, and one can still be important to the unfolding of the other but does not have to be.[4] There is no simple movement from place to place. Rather, because capital invested in the built environment flows from many investors acting in an uncoordinated fashion, because the built environment neither develops all at once nor deteriorates at the same rate and because some property investments require a prior disinvestment while others do not, investment, disinvestment and reinvestment can occur simultaneously.[5] If we focus our gaze on an area of a city at one point in time, we should find all three processes underway, even though one might be dominant.

Thus, using another metaphor from Neil Smith, to talk of a migrating 'frontier' between disinvestment and reinvestment and 'turning points' as one process overtakes another is to de-emphasize the simultaneous capital flows that define investment in the built environment.[6] The frontier metaphor captures only the dominant process within uneven spatial development, thereby making the equally important but less compelling processes less visible. In reality, the spatial demarcations among the processes of investment, disinvestment and reinvestment, I would argue, are generally not so sharp.

[3] For an introduction to the uneven development of the built environment see Harvey (1985a, pp. 62–89), Leitner and Sheppard (1989, pp. 67-77) and Smith (1979).

[4] Think of both ends of the see-saw being up or down at the same time. The metaphor no longer works.

[5] For an introduction to the complexity of the city building process, see Daunton (1990), Knox (1987), Warner (1962) and Weiss (1987).

[6] See Smith, Duncan and Reid (1989). An earlier and ecological approach to the issue is in Wright (1933).

To theorize properly notions of spatial simultaneity and spatial sequentiality, we must carefully delineate the geography of uneven development.[7] Are we speaking of a single site upon which a building or structure is being erected, demolished or under-maintained or of some larger area? Obviously, one can not demolish a factory and build a new factory on the same piece of ground at the same time, although one can certainly demolish a factory and build one adjacent to it on the same site or do so sequentially. Under-maintenance of a factory, moreover, is incompatible with its simultaneous upgrading. On a single site, then, one is more likely to find that capital investment is spatially sequential.

As one goes beyond a single property to multiple and contiguous properties within the same area, even as small as a city block, one is likely to discover greater spatial simultaneity. For example, an inner-city gentrifying neighborhood might well exhibit instances of disinvestment as factories and housing are abandoned; under-maintenance as elderly homeowners use their incomes for more pressing needs; and reinvestment as gentrifiers rehabilitate existing houses and developers clear sites for new housing or for small shopping areas.[8] Thus, while reinvestment in a gentrifying neighborhood might well dominate disinvestment and new investment, the three processes often exist there simultaneously. A similar argument might be made for central business districts and other areas of a city: a waterfront, for example.

In effect, because there are multiple investors with different interests and because property within an area neither is developed at the same time nor deteriorates at the same time, investment and reinvestment exist simultaneously in space. More precisely, the temporal and spatial uneven development of capitalism is both sequential and simultaneous, depending upon the scale at which one focuses and the time period of the analysis.

This is more than a debate about investment and disinvestment in the built environment. Our understanding of uneven development hinges on how we think about space and time within the overall dynamics of society. Edward Soja writes of a socio-spatial dialectic which represents '. . . a dialectically defined component of the social relations of production, relations which are simultaneously social and spatial.'[9] David Harvey has written that '[t]ime and space both get defined through the organization of social practices fundamental to commodity production' (Harvey, 1989, p. 239). Space is a social construction which unfolds temporally. Moreover, both agree that as technology

[7] The issue of spatial scale is crucial and has not received the theoretical and empirical attention it deserves. For an exception, see Smith and Dennis, 1987. This assessment also applies to sectoral switching, an issue not considered here but one which we have to integrate with spatial switching.

[8] For a discussion of an actual neighborhood experiencing these processes, see the analysis of Northern Liberties (Philadelphia) in Beauregard, 1990.

[9] See also Duncan, 1989.

minimizes temporal barriers and makes the production of space more central, time is subordinated to space in the current workings of capitalism. Neither Soja's nor Harvey's comments, however, should be viewed as precluding the possibility that space and time are constituted by a range of social practices which extend beyond commodity production and into cultural, social and political realms.

Uneven development, then, is ever-changing; it is transformed with every transformation of society itself. Moreover, uneven development is an expression and object of struggle among various forces of capitalist society; it is incessantly contested.[10] Any empirical analysis of uneven development, then, must expect to find complex patterns in the movement of capital across space and through time. These patterns are historically specific and embedded in the social practices that created them.

Philadelphia's housing market

Philadelphia is the fifth largest city in the country and a city whose postwar history has been dominated by an aggregate decline in population and employment (particularly in manufacturing), an expansion of its poor minority population and severe and large-scale deterioration. The 1980s were a time of gentrification adjacent to (and of robust office building development within) the central business district and a time of reinvestment along the waterfront and adjacent to the airport. Vast areas of the city, however, continued to decline physically and the city government teetered on the verge of fiscal bankruptcy. The city was suffering in aggregate terms, while sharp variations in development were indisputable (Beauregard, 1989a, 1989b, pp. 195–238).

Despite population loss, the city's housing market displayed a good deal of construction activity over the postwar period. From 1944 to 1989 nearly 210,000 units of new housing were constructed within the city, a yearly average of 4,565 units.[11] During roughly the same period (1950 to 1990), the population declined 26 percent, from 2.07 million to 1.54 million. When compared to the suburbs, though, the robustness of the city's housing market pales. From 1954 to 1989, approximately 21 percent of all metropolitan housing units were constructed in the city.[12] Only at two points in time—the early 1960s and again in the early 1980s—did city housing production essentially

[10] This point is essential to the Marxist perspective. See Sheppard and Barnes, 1986.

[11] These data are from building permit statistics made available by the Department of Licenses and Inspections, City of Philadelphia. Note that a building permit is an 'intent' to act, not a measure of the act itself.

[12] Data on metropolitan housing starts are from U. S. Department of Commerce, *Construction Review*, for selected years. More exactly, the data show 4,149 units per year for the city against a metropolitan average of 20,055.

equal suburban housing production. In the first period, city production was historically very high, while in the second period suburban production was historically low.

From a peak in 1961, new housing construction in the city has declined. The suburbs boomed between 1963 and 1975, but city construction rates fell from approximately 6,000 new housing units per year to approximately 2,000 units per year. By the 1980s, annual new housing construction in the city totaled about 1,800 units compared to about 6,700 units in the 1960s. Moreover, new housing construction in the city has been less volatile than that in the metropolitan area as a whole.[13]

If our evaluation of Philadelphia's housing market is based on long-term growth, competitiveness with the suburbs and the ability to sustain boom periods, then the city's housing market must be characterized as weak. On the other hand, if we assess it in relation to population growth, with population growth serving as a crude surrogate of demand, then our estimation changes. From this perspective, the city's housing market looks stronger. Population declined, but the number of households increased and new housing units continued to be built.[14]

The subject of this chapter, however, is not the overall performance of Philadelphia's housing market, but its temporal and spatial variability. To explore this, I propose to investigate the interaction of three flows of capital through that housing market: investment, disinvestment and reinvestment. The analysis begins with a temporal analysis of these flows and then turns to a more detailed look at their spatial array.

Temporal Uneven Development

The empirical basis of this analysis of uneven development is the epistemic correspondence between different theoretical movements of capital and different measures of housing market activity: investment measured by new housing construction, disinvestment measured by demolitions and reinvestment measured by conversions. The use of new housing units as an indicator of capital investment would seem to be non-controversial. The use of demolitions as a measure of disinvestment, however, is more problematic.

Demolition, the physical removal of a building, is only one form of disinvestment, arguably the endstate (that is, the last segment) of a more complex disinvestment process.[15] All capital has been taken out of the property and now the building only has

[13] From 1955 to 1978, the metropolitan area went through four residential building cycles, while the city went through two.

[14] Between 1960 and 1980, the number of households increased (615,767 to 619,781), as did the number of vacant housing units (22,320 to 65,350). Total housing units went from 649,033 to 685,629. Data are from the decennial Census reports on population and housing.

[15] For studies of disinvestment, decay and abandonment in inner-city housing markets see Lake, 1979; Marcuse,

(continued...)

to be physically removed. Yet, demolition in practice occurs even when a building has market value and is not functionally obsolete. Demolition might simply be a response to the site's rent gap or a politically motivated decision, as has been the case with urban renewal and highway construction. In addition, demolition requires an investment of sorts; the owner (or government) must pay to have it done.[16]

Conversion represents the restructuring of an existing investment and thus indicates reinvestment, as when an existing non-residential property is turned into a housing unit. Nonetheless, since almost any investment requires some conversion, even if it is only construction on unimproved land, the difference between investment and reinvestment is fuzzy. In general, reinvestment requires a significant upgrading of an existing building or structure; it involves rehabilitation or renovation. Investment refers to new construction—that is, the creation of a wholly new building or structure.[17]

Figure 3.1 displays data on the total number of new housing units added to the city housing stock and the number of units demolished.[18] At the time housing was being added in Philadelphia through new construction and conversions, a substantial amount of demolition was also taking place. Over the 1963–1989 time period, housing demolitions averaged 47 percent of total additions—1,537 versus 3,293 annually. Between 1963 and 1968, the two patterns converged as demolitions rose and total additions fell. Thereafter, demolitions continued to be a high proportion of total additions, reaching 50 percent on average after 1975. Conversions (not shown separately) averaged about 2 percent of new construction and about 4 percent of demolitions between 1963 and 1977, the years for which conversion data are available, and also exhibited a precipitous secular decline.[19]

[15](...continued)
1986; Salins, 1980; and Stegman, 1972. Abandonment, arson, code violations, tax delinquency and physical disrepair are alternative (but less accessible) measures of disinvestment.

[16] This poses an interesting measurement problem. The Philadelphia building permit files give neither the cost of demolition in terms of removal costs nor the value left within the building which is being destroyed. With new construction, all the costs of construction are included, excluding the cost of the land and architectural and engineering fees. For new construction, moreover, the sale price, of course, will be different. Thus the use of new construction and demolitions as measures of investment and disinvestment, respectively, do not represent accurately the relative capital flows of each process.

[17] Buildings can be upgraded, in the sense of becoming more valuable, in other ways, e.g. new leasing strategies for commercial buildings, new construction in the area of the property, greater demand for the area, which raises property values, and refinancing, which decreases debt relative to operating costs. In addition, new constructions, demolitions and conversions are affected by changes in nearby commercial and industrial areas, thus linking different property markets, an issue not discussed here.

[18] Data were collected from publications and files of the Department of Licenses and Inspections, City of Philadelphia. Conversions are not shown because such data exist only for 1961–1977 (with 1962 missing) and because they are so few in number that their visual presentation is unwieldy.

[19] There is always a problem of non-reporting for this measure. One can more easily hide a conversion from a building inspector than one can a demolition or new construction.

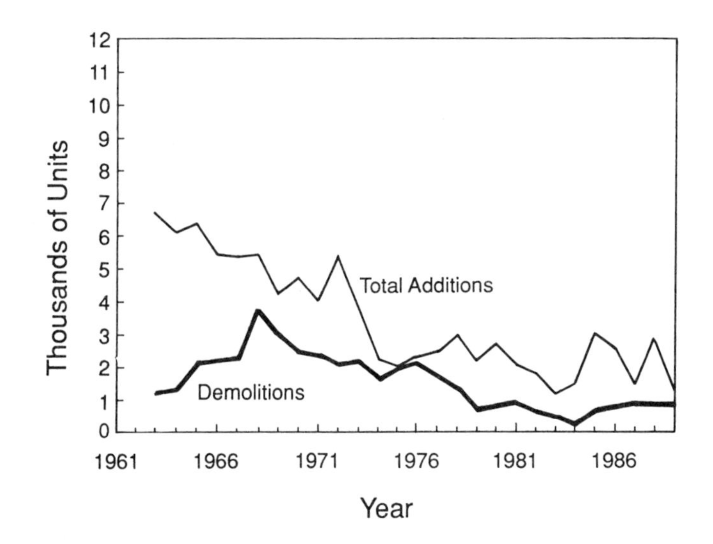

Fig. 3.1. Total housing additions and demolitions in Philadelphia, 1961–1989. *Note*: Total additions is the sum of new housing units constructed plus conversions. *Source*: Building permit data from the Department of Licenses and Inspections, City of Philadelphia.

Countercyclical behavior is evident but not overwhelming. Total additions hit a peak in 1961 and demolitions, a trough just two years later. Total additions had another peak in 1985 and demolitions, a trough in 1984. That countercyclical behavior, however, is not consistent across the three decades. New construction had a trough in 1975 and demolitions had a trough in 1974. At the new housing unit peaks in 1961, 1975 and 1985, demolitions were 22, 99 and 22 percent, respectively, of the total new housing units.

Overall, demolitions constitute a significant and mildly countercyclical contribution to housing activity within the city, though declining along with total additions. Reported conversions are minor compared to new construction and demolitions, though this ignores rehabilitation activity that does not involve conversions.[20] Thus the data indicate shrinking housing activity that is somewhat responsive to business cycles and a notable

[20] Rehabilitations provide a physical upgrading of an existing housing unit and are not always reported to authorities. The dollar value of rehabilitation activity is usually higher during downturns in the building cycle. See Beauregard 1991. Galster (1987) provides an extensive discussion of housing maintenance activity.

amount of disinvestment that is mildly countercyclical to new investment.

We can also explore the temporal patterning of these capital flows for subareas within the city, specifically the 12 'planning analysis sections' established by the city in the 1960s for statistical purposes (Fig. 3.2). Similar to the citywide findings, nine of the planning analysis sections had a decline in new construction over the twenty-five-year period from 1963 to 1986 (data not shown).[21] The three exceptions, all inner-ring areas, have neither decline nor growth. At the same time, the sections exhibit a great deal more volatility than the city as a whole. Total additions and, less so, demolitions rise and fall precipitously from one year to the next. Particularly striking as regards the long-run pattern is the relative dampening of almost all activity for many of the planning analysis sections beginning in the early 1970s.

Only four planning analysis sections exhibit any discernible countercyclical behavior between total additions and demolitions, and two of these are only very weakly countercyclical. Three sections display a cyclical pattern, with demolitions either lagging or leading total additions, and the remaining sections (with the exception of the Far Northeast) have a mixed temporal pattern, sometimes cyclical and sometimes countercyclical. The Far Northeast, because it had so few demolitions, simply cannot be described one way or the other. Thus, the individual planning analysis sections do not generally correspond to the citywide pattern of uneven temporal development. Rather, they are quite diverse.

Uneven Spatial Development

As for uneven spatial development, Figs. 3.2 and 3.3 present the distribution of total new housing units and total demolitions for the 1963–1986 time period across the planning analysis sections. The planning analysis sections, it should be noted, do not represent housing markets and generally comprise quite diverse neighborhoods. For example, Germantown/Chestnut Hill is a mix of middle- and lower-class row houses with some pockets of poverty (Germantown) and one of the city's most expensive neighborhoods (Chestnut Hill), which is made up of detached single-family houses on large lots.

Figure 3.2 shows new housing construction concentrated in Center City and the Far Northeast with 10,505 and 33,454 units, respectively, built in this time period, 52 percent of the citywide total (see also Table 3.1). At the other extreme, the southern parts of the city and Lower North and the Near Northeast had far fewer new units. The Southwest

[21] Since data on newly constructed housings units exist only for 1961–1986 and 1962 data are missing, the time period changes here.

is dominated by the international airport, a water treatment plant, swampland and industrial uses. The southern portion of South Philadelphia is also heavily industrial, although the northern portion was densely developed with housing and industry by the end of the nineteenth century. Upper North Philadelphia, Bridesburg/Kensington/Richmond, Germantown/Chestnut Hill and Olney/Oak Lane were generally built up prior to the 1960s, and the first two sections, along with Lower North Philadelphia, have the poorest

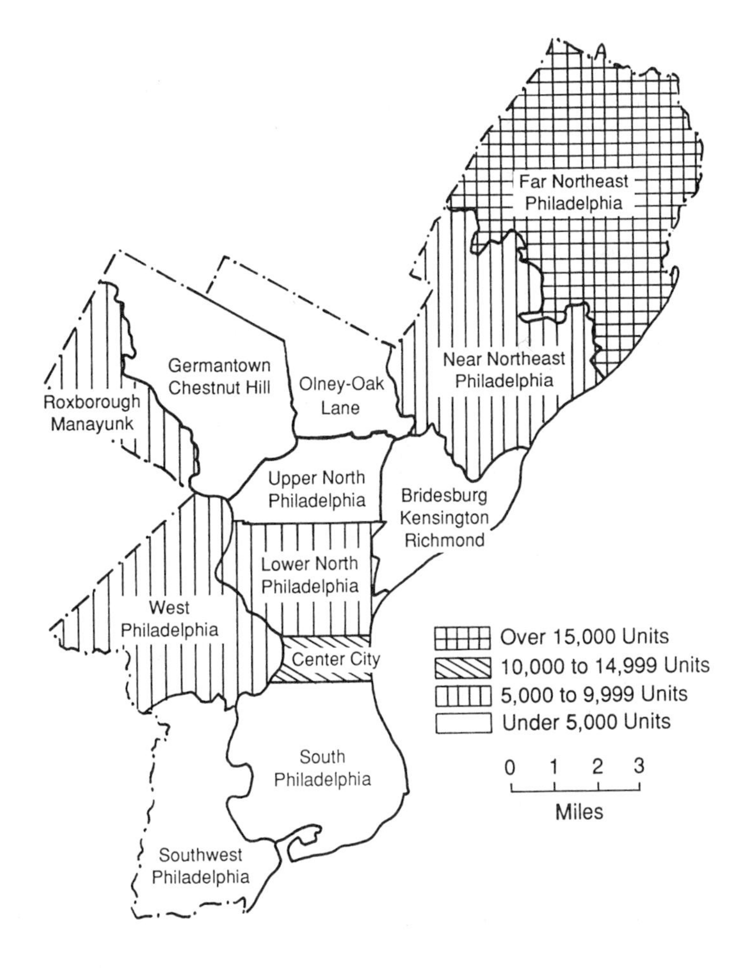

Fig. 3.2. New housing construction in Philadelphia by Planning Analysis Sections, 1963–1986.

populations in the city. In between, the Near Northeast was still developing in the early 1960s (though mainly in its northern areas). West Philadelphia, Lower North Philadelphia and Roxborough/Manayunk are not so easily explained, since all three were almost fully developed by the 1960s. West Philadelphia, though, is the location for major hospitals and universities and it, along with certain southern neighborhoods in Lower North Philadelphia, experienced gentrification in the 1980s.[22]

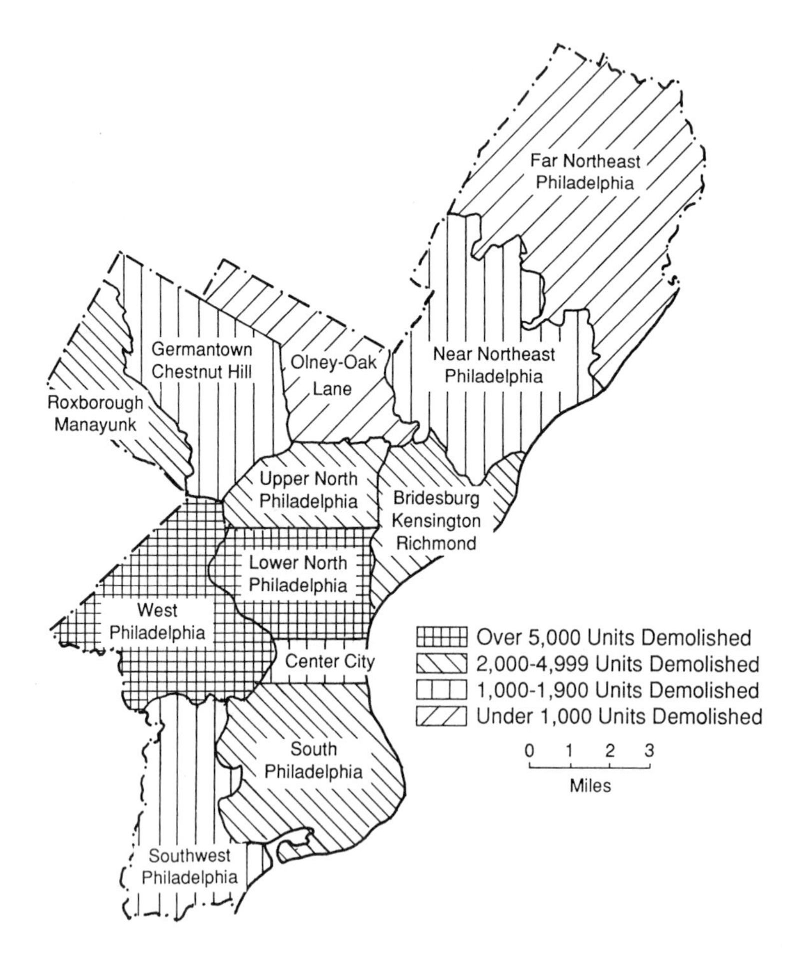

Fig. 3.3. Total demolitions in Philadelphia by Planning Analysis Sections, 1963–1986.

[22] Gentrification also occurred in the northern portions of South Philadelphia and the southwestern section of Center City.

Figure 3.3 presents data on demolitions by planning analysis section. The areas with the greatest numbers of demolitions are West Philadelphia and Lower North Philadelphia. Both are mixed-use areas, but the former has been an area of expanding non-profit institutions, while the latter has had a disproportionate share of abandoned factories and more publicly-financed slum clearance.[23] The next level of demolition occurred in South Philadelphia, Roxborough/Manayunk, Upper North Philadelphia and Bridesburg/Kensington/Richmond. The latter two have widespread industrial and residential blight, likely the most extensive in the city. Gentrification might have been the culprit in South Philadelphia, but the gentrification there was not necessarily dominated by new housing construction. Roxborough/Manayunk, again, is not easily explained. As for the areas with minimal demolitions, both had land upon which to expand after World War II.

Relatively few demolitions have occurred in Center City, particularly given the amount of new office construction in the 1980s, the relatively large number of new housing units constructed there in the last twenty-five years and the scarcity of vacant land. The land in this part of the city is very expensive and property values might have discouraged large-scale demolition. Moreover, much of the housing stock is historically significant. Combined with a well-developed local renovation industry, a tradition of restoration and a strong civic commitment to historic preservation, demolitions are understandably less numerous. In fact, the conversion data provide partial support for this contention; Center City dominated with 35 percent of all conversions and the Near and Far Northeast combined had 17 percent. The only other area with a large proportion of conversions was West Philadelphia, with 17 percent of the total. The fewest conversions took place in Roxborough/Manayunk, Lower and Upper North Philadelphia and Southwest Philadelphia.

Of course, any spatial distribution of housing activity in an older, industrial city like Philadelphia is likely to be affected by the spatial distribution of minorities and poor people. Minorities are often discriminated against in housing markets and are more likely to be of lower incomes than whites. The higher the level of poverty in an area, the fewer the new housing units that would be expected and possibly the demolitions.

In fact, five of the six areas with above-average African-American populations had below average new housing construction. The six areas with less than the average proportion of African-Americans were evenly split between above- and below-average new construction. In addition, areas with above-average proportions of African-Americans were more likely to have above average numbers of demolitions—four of the six areas. All areas below the citywide percentage of African-Americans had below-average

[23] On publicly-financed demolitions in Urban Renewal areas in Philadelphia, see Kleniewski (1986).

numbers of demolitions.[24]

The relation between poverty in an area and new construction was similar to that found between percent African-American and new construction (data not shown). Poverty and demolitons, moreover, are positively related; above-average poverty areas have above average demolitions and vice versa. In sum, areas with higher proportions of African-Americans and families living in poverty have had fewer new housing units constructed and more demolitions than areas with lower proportions.

Intersecting Capital Flows

To complete the analysis, however, we need to explore the spatial interaction of new construction, demolitions and conversions more closely. One way to do this is to cross-tabulate the data on the three dimensions. Essentially, we want to know if high (or low) amounts of new housing construction coexisted with high (or low) numbers of demolitions and conversions.

The data for the three measures and for each of the planning analysis sections are displayed in Table 3.1. One striking feature of these numbers is the broad range in the ratio of demolitions to new construction. In some areas, for every new housing unit added to the market, one or more (and even five in Bridesburg/Kensington/Richmond) units were subtracted from the housing stock. In contrast, other areas had 3, 4, 5 and up to 100 units (in the Far Northeast) of new construction for every one unit demolished. There is also spatial clustering. The areas with high demolition to new construction ratios are mainly in north Philadelphia and the areas with low ratios are in Center City and the newer northwest and northeast areas of the city.

Table 3.1 also includes the total constant dollar value of all new residential construction in these areas between 1963 and 1986. The association between numbers of new housing units and the dollar value of new residential construction is strong, as would be expected. Center City and the Near and Far Northeast are high on both, and Olney/Oak Lane, Upper North Philadelphia and Bridesburg/Kensington/Richmond are low on both. If one looks at the average value of newly constructed units (not shown), the relation between total dollar value and total units disintegrates, however. This is also the case for the relation between average value and demolitions. In effect, neither newly constructed units nor demolitions were concentrated in areas where new housing costs per

[24] The analysis is based on a 1970 measure of African-American presence. It represents an early year in the 1963–1986 time period to which subsequent housing activity is likely to have responded. This and the poverty measure discussed below (i.e., the percentage of families below the poverty line in 1969) are from the decennial Census population report for 1970.

unit were high or low.[25]

Table 3.1. Housing Market Activity, 1963–1986[a]

Planning Analysis Section	New Housing Units[b]	Demolitions[c]	Conversions[d]	Ratio[e]	New Construction Value[f]
Center City	10,505	1,725	418	0.16	121,372,547
South	2,809	3,454	97	1.23	26,210,015
Southwest	3,982	1,466	45	0.37	27,123,612
West	7,255	6,580	183	0.91	43,405,710
Lower North	4,975	16,002	52	3.22	47,814,141
Upper North	1,040	3,450	34	2.41	10,465,247
Bridesburg/ Kensington/ Richmond	382	2,087	27	5.46	3,560,421
Roxborough/ Manayunk	5,907	1,242	16	0.21	44,925,747
Germantown/ Chestnut Hill	4,464	1,289	94	0.29	46,353,994
Olney/ Oak Lane	2,414	349	37	0.14	14,280,541
Near Northeast	8,090	931	72	0.12	54,803,303
Far Northeast	33,454	365	130	0.01	254,081,352

Notes:
a. Data are totals for the time period.
b. Total number of new housing units constructed.
c. Total number of demolitions of housing units.
d. Total number of conversions to housing units.
e. Ratio of demolitions to new housing units.
f. Constant dollar (1958) value of new residential construction taken from building permits for the city.

[25] In 1987, the median sale price of housing in the city ranged from $155,500 in Center City to $9,390 in Upper North Philadelphia and averaged $37,000 citywide; see Warner (1989). Areas with low median sale price are also areas with proportions of families in poverty and of minorities.

Table 3.2. Relation of New Housing Construction to Demolitions For Planning Analysis Sections, 1963–1986

	Total New Housing Units [a]	
	Above Average	Below Average
Total Demolitions: [b] Above Average	West Philadelphia (0.91) [c]	South Philadelphia Lower North Upper North (2.29)
Below Average	Center City Near Northeast Far Northeast (0.10)	Southwest Bridesburg/Kensington Richmond Roxborough/Manayunk Germantown/Chestnut Hill Olney/Oak Lane (1.29)

Notes:
a. The section average for new housing unit construction is 7,106.4.
b. The section average for demolitions is 3,245.0.
c. Average ratio of total demolitions to new housing units for the sections in the cell. The citywide average is 1.21.

Table 3.2 shows the cross-tabular relation between new construction and demolitions for the 1963–1986 period. If there had been no relationship between the variables, we would find the planning analysis sections equally distributed across the cells. This is not the case. If demolitions and new construction were mutually exclusive, the planning analysis sections would cluster in the lower left and upper right corners. This is not the case either. In fact, a clustering exists in the lower right corner, where both demolitions and new construction are below the citywide averages. This might well be indicative of a city in aggregate decline, though that can not be said definitively. Still, only two planning analysis sections would have to move (from the lower right to the upper left) to produce 'no relationship.' Six would have to move to produce an inverse relationship. In effect, the relation between demolitions and new construction for Philadelphia in this period was closer to no relationship than to one in which they were mutually exclusive.

What about specifics? West Philadelphia, for example, offers the possibility that an above-average amount of new construction and demolition can co-exist spatially. Five areas have below-average new construction and below-average demolitions, indicating a housing market with very little activity. Overall, all four relationships are represented, though only one area is above average on both measures.

Note also that Table 3.2 displays in the respective cells the average ratios of demolitions to new construction for the planning analysis sections. Where there is below-average new construction, the average number of demolitions exceeds the average number of newly-built units, regardless of whether demolitions are above or below the average for the city. Where there is above-average new construction, the average number of demolitions falls below the number of new units. This suggests that where new construction is lacking, the housing stock is shrinking, not remaining stable. For those planning analysis sections below average on both indicators, however, the removal of one section (Bridesburg/Kensington/Richmond) would drop the ratio below 1.0 (to 0.25, in fact). Being below the citywide average on new construction does not have to result in a net loss to the housing stock, a possibility that tempers the relation between the lack of new construction and the shrinkage of the housing stock.

Though the data on conversions are not as extensive, we might still compare them to new units and demolitions. Using a three-way cross-tabulation, Table 3.3 displays these relationships using data only for 1963–1977, thus changing the yearly averages. This truncated time period partly contributes to the variation from Table 3.2.

Clearly, areas with below-average numbers of conversions are those with below-average numbers of new units and, within that category, with below-average numbers of demolitions. Only one area which was below-average on new construction and demolitions had an above-average number of conversions. Overall, conversions exhibit a strong and positive correlation with new construction and a weaker though still positive correlation with demolitions.

In sum, the data portray a city where approximately half of the housing submarkets had very little activity of any type. (Only one area, West Philadelphia, was above average on all three measures.) The table also suggests that the key is whether an area has below-average new construction activity. Where this has occurred, demolitions and conversions were below average. On the other hand, once new construction went above the citywide average, all combinations became possible. Finally, when the number of conversions was above the citywide average, above-average new construction and below-average demolitions tended to appear.

Table 3.3. New Housing Construction, Demolitions and Conversions in Philadelphia by Planning Analysis Sections, 1963–1977

	Total New Housing Units [a]			
	Above Average		Below Average	
	Total Demolitions [b]		**Total Demolitions**	
Total Conversions: [c]	Above Average	Below Average	Above Average	Below Average
Above Average	West (1.04) [d]	Center City Far Northeast (0.16)		Olney/Oak Lane (0.12)
Below Average	Lower North (1.14)	Near Northeast (0.09)	South (2.25)	Southwest Upper North Bridesburg/ Kensington/ Richmond Roxborough/ Manayunk Germantown/ Chestnut Hill (2.29)

Notes:

a. The section average is 5,412.7 units.
b. The section average is 2,891.8 units.
c. The section average is 100.4 units.
d. Numbers in parentheses are the average ratio of demolitions to new construction for the section. The citywide average is 0.50.

Assessing Demand

One of the possible criticisms of this analysis is that the data have not been controlled for the size of the housing submarkets and thus for the demand for housing within these planning analysis sections. Size, though, implies an understanding of the empirical and theoretical relationships between a scale factor (e.g., population) and the phenomenon being measured, in this case the amount of housing activity. In the instance of housing markets, the scale argument usually assumes that household formation, mediated by income, represents housing demand. Where new household formation is greatest, housing activity will also be greatest and vice versa. As a result, one needs to control for the change in size of the household population in order to identify accurately the spatial variations across the city and over time.

Evidence already presented provides some support for this argument. Total additions and demolitions have fallen as population has declined during the period under study, and the actual year-round occupied housing stock has grown, along with the number of households. Still, such a perspective ignores the relative autonomy of the city-building process and its potential to be driven by financial considerations only weakly associated with demographic changes (Haila, 1991). My caveat, however, does not negate the importance of assessing the possible scale effects within the data.

The simplest approach to weighing the demand for housing is to compare the total amount of new construction and demolition with the change in the number of households. Those data are not available for the exact time period under consideration and thus a surrogate measure is used—absolute change in the number of households between 1960 and 1980. Dividing the total amount of new construction and demolition in each planning analysis section by the net change in number of households living in those sections yields per capita measures which essentially control one aspect of housing demand.[26]

Table 3.4 shows the two-way cross-tabulation of the respective ratios for the twelve planning analysis sections. Note that six housing submarkets had net increases in the number of households over this time period. In each of these sections, the number of new units constructed per additional household exceeded the citywide average, and the number of demolitions per additional household fell below the citywide average. The remaining six sections had a net loss of households, and in each case new construction was below and the demolitions were above the citywide average. In sum, more new construction and less demolition occurred where the number of households was growing, and just the

[26] To repeat, effective demand (i.e., the financial ability of households) is another aspect, particularly in a city where one of the fastest-growing population segments over the postwar period has been lower-income households.

opposite was true for submarkets where the number of households was declining: less new construction, more demolitions.[27] This suggests that either housing activity was very responsive to changes in demographic housing demand or that migration was very responsive to housing market activity.

Compared to Table 3.2, where the data are not adjusted for changes in the size of the housing market, Table 3.4 shows less variation in the allocation of sections to cells. Controlling for scale strengthens the inverse relationship between new construction and demolitions. Nonetheless, it still presents a picture of a highly segmented citywide housing market: one segment of little new construction and significant demolitions and another of minor amounts of demolition and more variation in new construction activity.

Findings

Overall, the capital flows within Philadelphia's postwar housing market do not fall neatly into either a pattern of sequential or simultaneous uneven development. For the city as a whole, investment and disinvestment have behaved somewhat countercyclically; total additions and demolitions did not rise and fall simultaneously. This appeared against a background of long-term decline and weak performance relative to the city's suburban housing market. When the housing market was disaggregated spatially, the countercyclical relationship weakened. Only a minority of the submarkets were countercyclical; the remainder either exhibited cyclical relations or had no discernible temporal pattern. In addition, housing submarkets have been highly volatile and have experienced a long-term decline. In fact, a number of submarkets had almost no housing activity of any kind in the 1980s.

Spatially, investment in the city's housing market was highly uneven. The evidence partially supports a pattern of investment on the periphery and in the central business district and disinvestment in the neighborhoods and industrial areas surrounding the core where there are high concentrations of minorities and families in poverty. The connection among investment, disinvestment and reinvestment for these submarkets seems to be driven mainly by whether significant new construction is or is not occurring. Areas with above-average amounts of investment have either relatively high or relatively low amounts of disinvestment and reinvestment, but areas with little investment have minimal levels of disinvestment and reinvestment. For certain periods in the 1960s and early 1970s,

[27] In sections with an increase in households, 1.08 new units were built and 0.17 units demolished for each additional household. In sections with a decrease in households, 0.82 units were built and 0.64 units demolished for each net household lost.

however, submarkets with low amounts of investment often had a high degree of disinvestment.

Table 3.4. New Construction, Demolitions and Household Change in Philadelphia by Planning Analysis Sections, 1963–1986

Ratio of Demolitions to Household Change [b]	Ratio of New Construction to Household Change [a]: Above Average	Ratio of New Construction to Household Change [a]: Below Average
Above Average		South Southwest West Lower North Upper North Bridesburg/ Kensington/ Richmond
Below Average	Center City Roxborough/Manayunk Germantown/Chestnut Hill Olney/Oak Lane Near Northeast Far Northeast	

Notes:

a. The measure is the total number of new housing units for 1963–1986 divided by the absolute change in the number of households in the section between 1960 and 1980.

b. The measure is the total number of demolitions for 1963–1986 divided by the absolute change in the number of households in the section between 1960 and 1980.

The relationship between investment and disinvestment becomes much clearer when household demand is included in the analysis. Where demand has increased (that is, in about one-half of the submarkets), investment has been robust and disinvestment minimal. Just the opposite has been the case in the remainder of the city where the number of households has declined. Moreover, submarkets where households increased in number also had smaller proportions of minorities and families in poverty. Thus, there seems to be a link between investment and household formation, either through migration or the creation of new households from existing ones. The direction of causality, however, is not clear.

The analysis thus reveals a housing market that is spatially segmented along the dimensions of race and class. Where minorities and the poor are concentrated, disinvestment dominates capital flows, but through the 1980s neither disinvestment, investment nor reinvestment has been present to any significant degree. Throughout the remainder of the city, and despite long-term decline, investment continues, sometimes with reinvestment and sometimes without, sometimes with disinvestment and sometimes without. In spatial terms, then, capital flows tend towards simultaneity, though clearlyonly in one segment of the housing market. In temporal terms, and on a citywide basis, capital flows have tended to the countercyclical and thus to be sequential, though this is not the case for the city's submarkets. Segmentation, simultaneity and sequentiality characterize Philadelphia's postwar housing market.[28]

Through the twenty-five-year period of this analysis, then, a dual housing market has existed in Philadelphia. In almost one-half of the city, investment, disinvestment and reinvestment were all low. Little new housing was being built, few housing units were being demolished and conversions producing new housing units were rare. In this segment, the housing market has been essentially stagnant.

The other housing market segment in the city is more complex. It contains a relatively high level of new construction. The new units, however, do not always preclude a high degree of demolition activity or a paucity of conversions. Towards the outer limits of the city, where land existed for large-scale housing construction, few demolitions and conversions took place. In Center City, however, new housing construction and conversions were both prevalent, and in West Philadelphia extensive new construction, demolition and conversions coincided.

The findings of an uncertain countercyclical patterning and a spatially segmented housing market characterized by simultaneity and sequentiality, of course, might well be peculiar to Philadelphia during this time. Here was an older city experiencing the loss of

[28]Absent from this analysis is a breakdown of the 1963–1986 period into phases. This might reveal another set of relationships operating but poses a significant problem of periodization.

a middle-class white population and the influx of poor minorities, facing employment loss, competing poorly with its suburbs and having a large housing stock already in place but also areas available for new development. Such complex social and economic conditions, themselves expressed in uneven spatial and temporal patterns, are prime contributors to the complex capital flows that define the city's housing market.

Interpretation

For theorists, the causes and necessity of uneven development are pivotal issues around which to focus a geographic understanding of capitalism. Their arguments begin from the position that uneven development is endemic and then posit a broad pattern of spatial and temporal sequentiality. Investment and disinvestment see-saw across the landscape and through history, and the major capital flows are regional and global (Browett, 1984; Massey, 1984; Smith, 1984).

Leaving aside issues of cause and necessity, the purpose of this chapter has been to identify the fine-grained processes of uneven development at the urban scale. Neil Smith (1979; Smith *et al.*, 1989) has made important contributions to this understanding through his writings on the rent gap and frontiers of disinvestment and reinvestment. His concern is with the times and places where one capital flow exceeds another—for example, when a dominant process of investment is displaced by a process of disinvestment. The issue is one of temporal and spatial turning points, and the assumption is that once turning points appear, the newly dominant process persists and transforms its site until either overbuilding drives down profits or devaluation creates a rent gap (Badcock, 1989).

My objective has been to understand what takes place on either side of this frontier. I am more concerned with the weave of the urban fabric than with its edges and thus elevate to the center of the analysis the presumption that the processes of investment, disinvestment and reinvestment can occur simultaneously in space and time.

This weave, as the case study indicates, is not one in which capital flows are mutually exclusive. Rather, the processes occur together in complex patterns. In places experiencing investment booms, investment does overwhelm disinvestment. At the other extreme, where disinvestment reigns, investment and reinvestment might or might not appear. In fact, if there is a continuum with investment at the one end, the other end is not large-scale disinvestment but stagnation—places where neither investment, disinvestment nor reinvestment are present. Major disinvestment might well precede stagnation, thereby creating places where capital flows are virtually non-existent.

The majority of places fall between the three extremes. There, the weave of capital flows is highly complex, sometimes sequential, sometimes simultaneous and sometimes an unstable mixture of the two. Nonetheless, disinvestment and reinvestment are always present, even if only at low levels. If one had to choose between whether uneven development is sequential or simultaneous, these places suggest that simultaneity is the more appropriate choice.

While the evidence presented to support that conclusion cannot tell us why such complex patterns exist, we do know enough about the city-building process to sketch a possible answer. That answer would consist of four elements. First, it would note that any investment in or disinvestment from the built environment of cities faces a historical layering of buildings and structures, set forth in complex spatial patterns, that fracture those processes (Beauregard, 1989c, pp. 209–240). Owners of buildings, existing mosaiclike in the landscape, neither invest nor disinvest at similar rates. Second, multiple owners are joined by multiple investors and numerous other participants (e.g., builders, real estate brokers, zoning officers) in the city building process, all of whom have diverse motives and varied capabilities. In turn, their distinct behaviors are also fostered by the differing dynamics of sectors (e.g., industrial, commercial, residential, renovation) within the construction industry. None of these groups responds simply to signals sent from the sphere of production. Capital flows into and through the built environment are relatively autonomous from decisions about the consequences of capital investment in economic development.

Third, one cannot understand the uneven development of the built environment solely in terms of capital flows, as if the built environment were a commodity lacking in anything but exchange value. For example, the economistic housing market is 'distorted' by issues of race and class. Additionally, housing activity in a neighborhood (particularly reinvestment) is very much affected by perceptions, expectations, marketing schemes, neighborhood organizations and other practices driven by more than financial considerations. Uneven development is a social as well as an economic phenomenon.

Finally, an explanation for uneven development, particularly for the postwar period, should attend to the role of governments in shaping the opportunities within and the conditions of the built environment. The impact of FHA mortgage insurance on privileging suburban home ownership over inner-city living, the dire consequences for inner-city neighborhoods of interstate highways, and the devastation wrought by Urban Renewal's massive slum clearance are well known. The uneven development of the urban fabric is not without political dynamics.

In addition, this case study of the capital flows within Philadelphia's postwar housing market has traits which might become important in developing a broader explanation for uneven development.[29] Over this period, Philadelphia was a city in aggregate decline;

[29] Epistemologically, I believe this case captures the workings of capitalism. Therefore, other cases should reveal (continued...)

analyses of growth areas might yield much different patterns. Second, this is a study of a city, not a region. The flows of capital have been analytically confined to spaces within the city's boundaries. Arguably, housing markets are really metropolitan in scale, particularly when one considers financing practices. Moreover, by focusing solely on the city, investment opportunities and flows of capital originating outside the city were excluded. Third, the subject here is housing. To the extent that capital crosses different building types (e.g., residential, commercial, industrial) and flows from one circuit of capital to another (e.g., from the built environment to the sphere of production), this chapter presents a partial view of uneven development. Finally, the time period chosen was dictated by the availability of the data rather than by a theoretical analysis of turning points in the building of Philadelphia. When weighing the chapter's contribution to the interpretation of uneven development and contemplating future research, serious attention needs to be given to each of these caveats.

While my findings can be likened more to a paint-splattered Jackson Pollock than a crisp and linear Mondrian, one implication is clear. Growth and decline, renewal and blight, expansion and contraction are not mutually exclusive phenomena, whether viewed along the axis of time or the axis of space. Rather, processes of development and underdevelopment exist simultaneously. The world is not one or the other; it is dialectically both and incessantly turbulent.

Acknowledgments

The analysis is based upon work supported by the National Science Foundation under Grant No. SES 86-07890. Linda Dottor and Jun Hyun Hong provided research assistance, and Jim Weiss of the Department of Licenses and Inspections, City of Philadelphia, was an invaluable resource. Thanks to Ira Goldstein, Paul Knox, Neil Smith and Daphne Spain, who gave close reads to an earlier version and offered numerous and useful criticisms. Thanks also to the participants in a colloquium at West Virginia University in March 1991 for their helpful comments.

[29](...continued)

a roughly similar 'logic' in operation. That is, the findings are ripe for theoretical generalization if not statistical generalization. Still, capitalism and uneven development appear in many guises, and one cannot understand them by focusing on one spatial scale or a single time period.

References

Badcock, B. 1989 An Australian View of the Rent Gap Hypothesis, *Annals of the Association of American Geographers*, 79, 125–45.

Beauregard, R. A. 1991 Capital Restructuring and the New Built Environment of Global Cities: New York and Los Angeles, *International Journal of Urban and Regional Research*, 15, 90–105.

Beauregard, R. A. 1990 Trajectories of Neighborhood Change: The Case of Gentrification, *Environment and Planning A*, 22, 855–74.

Beauregard, R. A. 1989a City Profile: Philadelphia, *Cities*, 6, 300–308.

Beauregard, R. A. 1989b The Spatial Transformation of Postwar Philadelphia, in R. A. Beauregard (ed.), *Atop the Urban Hierarchy*. Totowa, NJ: Rowman and Littlefield.

Beauregard, R. A. 1989c Space, Time, and Economic Restructuring, in R. A. Beauregard (ed.), *Economic Restructuring and Political Response*. Newbury Park, CA: Sage Publications.

Bradford, C. P. and Rubinovitz, L. S. 1975 The Urban-Suburban Investment-Disinvestment Process, *Annals of the American Academy of Political and Social Science*, 422, 77–86.

Browett, J. 1984 On the Necessity and Inevitability of Uneven Spatial Development Under Capitalism, *International Journal of Urban and Regional Research*, 8, 155–76.

Clark, G. L. 1980 Capitalism and Regional Inequality, *Annals of the Association of American Geographers*, 70, 226–37.

Daunton, M. J. 1990 *Housing the Workers*, London: Leicester University Press.

Duncan, S. S. 1989 Uneven Development and the Difference That Space Makes, *GeoForum*, 20, 131–39.

Galster, G. 1987 *Homeowners and Neighborhood Reinvestment*. Durham, NC: Duke University Press.

Haila, A. 1991 Four Types of Investment in Land and Property, *International Journal of Urban and Regional Research*, 15, 343-65.

Harvey, D. 1989 *The Condition of Postmodernity*. Oxford: Basil Blackwell.

Harvey, D. 1985a *The Urbanization of Capital*. Baltimore: The Johns Hopkins University Press.

Harvey, D. 1985b *Consciousness and the Urban Experience*. Baltimore: The Johns Hopkins University Press.

Harvey, D. 1982 *The Limits to Capital*. Chicago: University of Chicago Press.

Kleniewski, N. 1986 Triage and Urban Planning: A Case Study of Philadelphia. *International Journal of Urban and Regional Research*, 10, 563–79.

Knox, P. 1987 The Social Production of the Built Environment, *Progress in Human Geography*, 11, 354–77.

Lake, R. W. 1979 *Real Estate Tax Delinquency*. New Brunswick, NJ: Center for Urban Policy Research.

Leitner, H. and Sheppard, E. 1989 The City as Locus of Production, in R. Peet and N. Thrift (eds.), *New Models of the City*. London: Unwin Hyman.

Marcuse, P. 1986 Abandonment, Gentrification and Displacement, in N. Smith and P. Williams (eds.), *Gentrification of the City*. Boston: Allen & Unwin.

Massey, D. 1984 *Spatial Divisions of Labor*. New York: Methuen.

Myrdal, G. 1957 *Economic Theory and Underdeveloped Regions*. New York: Harper and Row.

Richardson, H. W. 1969 *Regional Economics*. New York: Praeger.

Richardson, H. W. and Turk, J. H. (eds.) 1985 *Economic Prospects for the Northeast*. Philadelphia: Temple University Press.

Salins, P. D. 1980 *The Ecology of Housing Destruction*. New York: New York University Press.

Schumpeter, J. A. 1950 *Capitalism, Socialism and Democracy*. New York: Harper and Row.

Scott, A. and Storper, M. (eds.) 1986 *Production, Work, Territory*. Boston: Allen & Unwin.

Sheppard, E. and Barnes, T. 1986 Instabilities in the Geography of Capitalist Production, *Annals of the Association of American Geographers*, 76, 493–507.

Smith, N. 1984 *Uneven Development*. Oxford: Basil Blackwell.

Smith, N. 1979 Toward a Theory of Gentrification, *Journal of the American Planning Association*, 45, 538–48.

Smith, N. and Dennis, W. 1987 The Restructuring of Geographical Scale, *Economic Geography*, 63, 160–82.

Smith, N., Duncan, B. and L. Reid. 1989 From Disinvestment to Reinvestment: Tax Arrears and Turning Points in the East Village, *Housing Studies*, 4, 238–52.

Soja, E. W. 1989 *Postmodern Geographies*. London: Verso.

Stegman, M. A. 1972 *Housing Investment in the Inner City: The Dynamics of Decay*. Cambridge, MA: MIT Press.

Stillwell, F. J. B. 1978 Competing Analyses of the Spatial Aspects of Capitalist Development, *The Review of Radical Political Economics*, 3, 18–27.

Thompson, W. R. 1968 *A Preface to Urban Economics*. Baltimore: The Johns Hopkins University Press.

Warner, S. 1989 Philadelphia Planners Find Home Prices Climbing, *The Philadelphia Inquirer*, January 8, Section H, p. 1.

Warner, S. B. 1962 *Streetcar Suburbs.* Cambridge, MA: Harvard University Press.

Watkins, A. J. 1980 *The Practice of Urban Economics.* Beverly Hills, CA: Sage Publications.

Weiss, M. 1987 *The Rise of the Community Builders.* New York: Columbia University Press.

Wright, H. 1933 Sinking Slums, *Survey Graphics*, 22, 417–19.

Four

IDENTITY AND DIFFERENCE: THE INTERNATIONALIZATION OF CAPITAL AND THE GLOBALIZATION OF CULTURE

Anthony D. King

State University of New York at Binghampton

Of the paradigm shifts that have taken place in the study of the city in the last two decades, few can be more significant than the change in the choice of matrix within which the city is seen to be embedded. Whereas scholars in the 1970s generally accepted the nationally-defined society as the appropriate unit, burgeoning evidence of the globalization of production, massive international labor migration, satellite communication and other factors have now made it clear that the relevant framework is international and global.

This realization has had various effects on urban research. First, a growing number of studies are charting the links of particular cities to the world-economy, including policy-oriented projects on institutions in cities which increasingly position themselves globally (Henderson and Castells, 1987; Knight, 1989; LPAC, 1990; Sassen, 1991a). Second, what scholars see as the contemporary reality of the world-economy, the world-system or globality (the distinctions are discussed below) has encouraged the rehistoricizing of particular urban issues and sites in relation to the development of the world-economy and processes of globalization (King, 1990a, 1990b). Third, given the susceptibility of specific nations, regions or cities to changing fortunes in the world-system, attempts to identify differences in their economic, social or political histories by reference to distinctive cultural or geographical characteristics has put a new emphasis on comparative research as well as on studies of localities (Cooke, 1989).

Each genre of studies presupposes a conceptualization, a perspective rather than a theory or model, of that larger whole within which the city can be located. The result is a growing interest in the different theoretical conceptualizations of 'the world as a whole' and the methodological approaches they imply, which scholars see as relevant for investigating different dimensions of urban social, spatial or cultural reality. The clearest example of this is illustrated by work on the world or global city.

The term 'world city' has been around for some time, yet the meanings it signifies have considerably changed. Already in 1889, Goethe applied the term *Weltstadt* to Rome and Paris, apparently referring to the special cultural eminence of these two cities (Gottman, 1989, p. 62). In 1915, Patrick Geddes shifted the frame of reference to a more economic and commercial context: 'world cities' were those 'in which a disproportionate part of the world's business is conducted' (Geddes, 1949, p. 2).

Some fifty years later, Peter Hall, while accepting Geddes' basic definition, added a political dimension. World cities were 'the major centers of political power, the seats of the most powerful governments, international authorities, of trade, etc.' (Hall, 1984, pp. 1–2); for Braudel, the world city was a center of specific world-economies (1984, p. 26).

The major shift in understanding, however, occurred in the 1980s and it came about as much from the reconceptualizations of the matrix and whether this is being represented as a world economy, a world-system, an international or world society, a global culture or a new international division of labor, among others, as from changes in the concept of the city itself. Seen from the 'outside' of a national society or from certain privileged positions within it, questions of economic development prioritize structures of the world-economy, but 'internally-defined' issues of cultural identity, ethnic nationalism, religious fundamentalism and their manifestations are not easily contained within these concepts. To date, most work has been directed at cities in the world economy.

Charting the international spread of manufacturing, with trade occurring more frequently between subsidiaries or joint ventures of global corporations producing goods in different parts of the world, Cohen (1981) drew attention to the international spread of corporate-related services, including multinational banks, law, accounting, advertising and contracting firms. Along with these developments were the international financial markets. Premised on these developments, he identified the emergence of a hierarchy of global cities, 'international centers for business decision-making and corporate strategy formulation. In a broader sense, these places have emerged as cities for the coordination and control of the new international division of labor' (p. 300).

Subsequent formulations have articulated this argument with increasing strength. Friedman and Wolff (1982) write about 'the spatial articulation of the emerging world system of production through a global network of cities,' their interest being in 'the principal urban regions in this network in which most of the world's active capital comes to be concentrated, regions which play a vital part in the great capitalist undertaking to organize the world for the efficient extraction of surplus.' Sassen-Koob refers to 'global control centers . . . nodal points to coordinate and control global economic activity' (1986, p. 88; 1991) where are located the headquarters of the major banks, the multinational corporations and centers of ideological control. According to Smith and Feagin (1987, p. 6) major cities tend to specialize in particular aspects of raw materials production, distribution, financial and other service activities. Berry (1990) sees the world economy as being controlled effectively from 19 urban regions within which all the transnational corporation headquarters are located, 10 in North America, 6 in Europe and 3 in Asia, mainly in Japan and Korea. Yet, as Smith and Feagin point out, it has become commonplace to note the global context of urban economies, but clarity about what this global context means is much less common. What does seem clear is that a distinction needs to be made between cities which have a dominating role in global transactions and others which simply participate in transactions which are worldwide (Jones, 1991).

This dominating role has been especially brought out by Sassen (1991a), who sees a hierarchy of global cities emerging which constitute a system rather than competing with each other. More importantly, she sees the possibility of a 'systematic discontinuity' between what used to be thought of as national growth and the forms of growth evident in global cities in the 1980s. In other words, the economic links of these cities to each other, within the world-economy, are more important than to the nation state within which each exists. This interesting proposition has an immediate parallel in the notion of a putative global urban culture (King, 1991b), where specific aspects of world cities, and the cultures (or subcultures) within them, are seen to have more in common with each other than with the subcultures of the state where the city exists.

These formulations, however, suggest that it is primarily the economic and commercial function of the city that is being economy foregrounded in these accounts. Yet this emphasis both neglects other dimensions of a city's global significance and also marginalizes the roles of other cities in the political, ideological and cultural/religious spheres. It also underestimates the ideological work which architecture, space and the built environment in general perform in representing the interests of capital on a global scale (e.g., in New York or Rio de Janeiro) or the way the changing ethnic, racial or religious composition of cities affects their cultural politics and cultural production with potentially significant local, national and global effects, not least on their economic role (Algiers, Beirut, Tehran, Washington). It also marginalizes the importance of social, cultural or political movements in regions outside these world-city centers.

Theorizing globality

How this world or global matrix is defined and conceptualized, however, depends on the particular focus of research. Only in an unmediated and undifferentiated way can we say that there are 'cities in a global society' (Knight and Gappert, 1989). Dependent on whether our interest is in structural changes in international banking, the transformation of cultures or the forms of the built environment, the 'world-level' framework can be described in many different ways: the international system, the international division of labor, global society,[1] humankind, East-West/North-South, First, Second and Third World, etc.

Some of these conceptualizations are as old as the world religions which used them; others, like world-economy and the international division of labor, date from the last third of the nineteenth century. Still others, like First/Second/Third World, are as recent as the early 1950s (Wolff-Phillips, 1987). As the intention of this chapter is to focus on the economic, social, cultural and built-environment transformations in cities as a result of international and global processes, I shall first examine five recent conceptualizations of 'the world' or 'globality' which I believe useful for understanding these. The first is the concept of the world-system.

The world-system perspective

Wallerstein's notion of the world-system emerged as a direct outcome of the abandonment of the sovereign state or national society as the unit of analysis: 'neither one was a social system and . . . one could only speak of social change in social systems' (1974, p. 8). Wallerstein argues that the basic unit of analysis should be the 'historical system' of which the boundaries are 'those within which the system and the people within it are regularly reproduced by some kind of ongoing division of labor' (So, 1990).

The modern world-system 'took the form of a capitalist world-economy that had its genesis in Europe in the long sixteenth century and that involved the transformation of a particular redistributive or tributary mode of production, that of feudal Europe . . . into a qualitatively different social system. Since that time, the capitalist world-economy has (a) geographically expanded to cover the whole globe; (b) manifested a cyclical pattern of expansion and contraction . . . and (c) undergone a process of secular transformation including technological advance, industrialization, proletarianization and the emergence of structured political resistance to the system itself—a process that is still going on today' (Wallerstein, 1980, pp. 7–8). The capitalist world-system is presented as consisting of the core, semi-periphery and periphery, world regions which are primarily economic sectors to which capitalists transfer capital, sectors which profit from the wage

[1] International denotes the primacy of, and relations between, nation states; and transnational denotes processes which go across them. Global processes develop either irrespective of interventions of the state or with its concurrence, in realms of activity which occur by accident or default (such as global warming) or conscious design (global communications, global trading) on the part of institutions, individuals or governments. Used in this oppositional sense to inter- or transnational, global implies the real or potential existence of a phenomenon in every state and every region of the territorial globe. National, international, etc., classifies people, phenomena and events as national subjects/objects, locating them principally in relation to the nation state. Global/globalization does not deny these classifications but suggests the possibility of other categories of identity and association.

productivity squeeze in the leading sectors (So, 1990, from which the following is taken).

The emergence of the contemporary capitalist world-economy resulted from the extraction of surplus into the core from what became the peripheral zones, with these geographical zones changing over time. The history of the world-system is of a gradual incorporation of countries and peoples into a single capitalist world-economy.

Despite the triumphs of anti-systemic movements (social democratic labor movements and socialist countries), the capitalist world-economy has steadily expanded, particularly since 1945. Studies in the world-system perspective focus especially on the secular trends (incorporation, the commercialization of agriculture, industrialization, proletarianization) and the cyclical rhythms of expansion and contraction. Thus, in a downward phase of the world-economy, the core weakens hold over the periphery; in the upward phase, the reverse will happen (p. 197).

Apart from earlier critiques (*American Journal of Sociology*, 1975), the controversies surrounding the world-system perspective have recently been spelled out at some length (So, 1990; Frank, 1991; Wallerstein, 1991a). Among the strengths are its theoretical structure, the merit of treating the world as the unit of analysis and the broad research focus which concentrates both on core and periphery and the relations between them. The principal criticisms are directed at the tendency for the world-system to take on a life of its own so that specific historical events at national and subnational levels are neglected, becoming subordinate to it. Other comments draw attention to the lack of emphasis on class relations within nations that shape the global relations between them, the underestimating of agency at the expense of structure, and a supposedly Eurocentric perspective (Frank, 1991). Lack of attention to the sphere of culture (Boyne, 1990) is countered in Wallerstein's recent work (1991b). If these comments are noted, looking at cities, urbanization and the built environment from a world-system perspective has various interesting possibilities.

While the language of 'core-periphery' may seem to take on its own level of determinacy, marginalizing the input of local societies and cultures (Abou-El-Haj, 1991), it nonetheless conveys the modes of economic and political power at the global level. Also worth mentioning is the emphasis it places on the interrelatedness and systemic development of economies. The merit of the perspective is more in terms of its long historical range, broad geographical scope, the insights it gives into the global inter-connectedness of cities and, especially, the rise of particular port cities, as regions are incorporated into the capitalist world-economy. Historically, the existence in urban settlements at either end of an emerging international division of labor, of specific institutions and building forms, with particular economic, social or political functions, and architectural styles, can be understood from this perspective (King, 1990a). As global inter-connectedness becomes increasingly apparent from the nineteenth century, world-system ideas increase their salience; the problems arise with applying them ex post facto and too uncritically to the past.

The system of world cities and its hierarchical ordering within a global system of production are presaged on tenets of the world-system perspective. As both Smith and Feagin (1987) and Berry (1990) point out, the central economic actors in the international division of labor are the top 500–1000 multinational corporations, whose headquarters' facilities are disproportionately located in the major cities of core countries.

Globalization

Where Wallerstein prioritizes the role of capital in structuring the capitalist world-economy into three sectors of uneven development which change over time, Robertson sees capitalism as only one dimension of 'globalization,' which he defines as 'the process by which the world becomes a single place' and 'the consciousness of the globe as such' (Robertson, 1987a). Identifying the four main components in the process of globalization as national societies, the system of international relations and conceptions of individuals and of humankind, Robertson (1990) traces out five phases of the temporal-historical path leading to the very high degree of global density and complexity at the end of the twentieth century. For example, the first phase in Europe, from the early 15th to the mid-18th century, sees the incipient growth of national communities and the downplaying of the medieval transnational system, the heliocentric theory of the world and the spread of the Gregorian calendar.

Phase two, in Europe, from the mid-18th to the late-19th century, sees shifts towards the idea of the homogenous unitary state, formalized international relations and the thematization of the nationalism-internationalization issue. The third phase, late-19th to the mid-20th century, shows international formalization, the attempted implementation of ideas about humanity, the increase in the number of global competitions (e.g., the Olympics, Nobel Prizes) and the implementation of World Time. Phase Four sees globewide international conflicts concerning forms of life, the Holocaust, Atomic Bomb, United Nations, and Phase Five, the 'uncertainty phase' (to the 1990s) emphasizes heightened global consciousness (ecology, health, humanitarian concerns), an increasing number of global institutions and movements, societies faced with problems of multiculturality and polyethnicity and gender, race and ethnicity making conceptions of the individual more complex. Elsewhere, Robertson states that the contemporary concern with civilizational, societal and ethnic uniqueness, expressed through such themes as identity, tradition and indiginization, largely rests on globally diffused ideas; in an increasingly globalized world, there is an exacerbation of societal and ethnic self-consciousness (Robertson, 1987a). There is, in Robertson's view (1990), a temptation to account for the present state of globality by reference to one particular factor or process, e.g., 'Westernization,' 'imperialism,' 'Americanization.' Robertson argues for the analytical separation of the factors which have led to a single world—the spread of capitalism and Western imperialism, the development of a global media system, etc.. Commenting on some of these ideas, Featherstone (1990, p. 11) suggests that there is little prospect of a global culture; rather, there are global cultures in the plural. Moreover, discussion about global culture is generated at a particular time, in a particular place and within a particular language and discourse.

While the strengths of globalization theory are in the unitary nature of the perspective as compared, for example, to the multiplicity of internationalism and the tripartite classification of the world-system perspective, its weakness is in not giving sufficient recognition to global inequalities of resources and power. So-called 'global' institutions, economic, financial, cultural, and ideological, are overwhelmingly 'Western,' located in, or emanating from, Western cities. Moreover, the theory itself (as with all conceptions of globality, world systems, etc.) is centered in Euro-America, prefaced on a European concept of time and history, though Robertson is indeed concerned with how other cultures and civilizations represent the world as a whole.

Nonetheless, Robertson's globalization theories provide seminal suggestions for understanding both urban and building development. Historical phases of the globalization process are inscribed in cities in many different ways, not only in the movements of international capital but in the transplantation of ideas, institutions and cultural practices. Singular international institutions of an economic and political nature (the World Bank in Washington, the United Nations in New York, the International Court of Human Justice in The Hague) or global cultural/religious institutions (the Qabah, Mecca, or St. Peter's, Rome), have, in the last two decades, been supplemented by recurring global cultural events (World Cup, World Games, International Conventions, pop and media festivals), which are being used to transform the form, nature and economy of cities. For example, the substantial increase in the number of international conventions in the 1980s has greatly increased the size, scale and number of hotels and convention halls in many cities (Table 4.1).

In spatializing Robertson's notion of globalization, it is clear that this is not a homogenous process but one filtered through and affected by history, culture, political power, geography, shifts in capital and the policies of individual states. Thus, while multiculturality and polyethnicity are increasingly manifest in many societies (discussed below in relation to New York and London), the way they get represented in the built environment is subject to the rules, codes and policies of individual places and states. Moreover, multiculturality and polyethnicity combine different cultures, different ethnicities, different histories. If capital has its own logic (the logic of central-city land values and the profit to be reaped from suburban development), local states and cultures also make their unique impact.

Postcolonialism

The specificity of both the social and cultural as well as the architectural and physical-spatial characteristics of many world cities can also be understood by reference to two other world-level paradigms: post-colonialism and post-imperialism. For varying periods, ranging from decades to centuries, some seven-eighths of the world was subject to different forms of European, American and Asian colonialisms. Many institutions and practices, from education and language to urban planning and government, have been formed and transformed by colonialism and imperialism: cities have been built as the direct result of these phenomena (King, 1990a).

Post-colonialism refers to the social, demographic, political, cultural, and spatial and built form conditions in once-colonial societies of the periphery; post-imperialism, to these phenomena in what were once the metropolitan capitals and heartlands at the imperial core.

Where the emphasis of the world-system perspective is on the overriding political and economic trajectory of the colonial process (often, as in Latin and Central America, subordinating the ideological and religious motive of the colonial enterprise to economic aims) and theories of globalization have not yet spelled out the historical specificities of this process, colonial and post-colonial paradigms address issues which, while recognizing economic motivations, also focus on structural features common to colonial situations, arising from questions of class, gender, ethnicity, religion and race and their representation in the built environment. Questions of political and cultural power and their translation and symbolic representation are also addressed, as are the distinctive themes of hybridization, syncretization and cultural transformation. Unlike the generalizations of the modernist and postmodernist paradigm (see below), studies of colonialism and postcolonialism have real political and historical referents in space and time, locating cultural as well as economic and political connections between metropole and colony.

The postcolonial paradigm recognizes a range of characteristics often inherent in postcolonial cities, though the extent to which they are found is dependent on state policies; for example, the macrocephalous port city, often the country's capital, and the persistence of export-oriented development; the drive to develop new inland capitals; the simultaneous existence of three, four or more different forms of built environment (including modes of construction and materials, financing, use, etc.) resulting from three, four or more historic modes of production: peasant agriculture, colonialism, state socialism

Table 4.1. Changes in the Number of International Conventions Classified by City, 1982–1986*

	City Rank / Conventions Held									
	1986		1985		1984		1983		1982	
City	**Rank**	**#**	**Rank**	**#**	**Rank**	**#**	**Rank**	**#**	**Rank**	**#**
Paris	1	358	1	274	1	254	1	252	1	292
London	2	258	2	238	2	248	2	235	2	242
Geneva	3	180	4	212	4	175	3	153	3	147
Brussels	4	157	3	219	3	201	4	145	4	118
Madrid	5	118	27	37	31	31	15	51	33	22
Vienna	6	106	5	127	5	146	5	142	5	90
W. Berlin	7	100	6	94	8	76	12	62	11	47
Singapore	8	100	10	74	10	68	6	77	14	44
Barcelona	9	96	13	63	32	30	37	21	—	8
Amsterdam	10	84	19	47	12	64	18	44	19	36
Seoul	11	84	12	65	17	47	35	24	19	36
Wash., D.C.	12	75	16	54	15	53	17	45	13	45
New York	13	72	8	90	7	84	9	65	6	70
Rome	14	69	7	91	6	85	7	73	8	69
Strasbourg	15	67	9	80	9	74	11	64	10	52
Munich	16	63	13	62	17	4	23	36	23	34
Copenhagen	17	63	11	71	13	62	8	72	6	70
Stockholm	18	63	15	60	21	43	21	39	27	28
Tokyo	19	56	17	53	14	54	13	55	9	55
Hong Kong	20	54	21	44	20	45	14	52	12	46
Budapest	21	53	17	53	19	46	9	65	18	37
Helsinki	22	52	24	41	11	65	28	28	23	34
Montreal	23	50	20	45	23	37	18	44	30	24
Bangkok	24	47	30	35	24	36	29	27	19	36
Buenos Aires	25	46	38	28	32	32	36	22	36	20
Total		6,681		6,163		5,795		4,864		4,353

*It is worth noting that of these 25 cities, 16 are in Europe, 5 in Asia, 3 in North America and 1 in South America. From Knight and Gappert, 1989, p. 323.

and global capitalism; the emphasis of the state on national cultural agendas, contested by different class, regional, religious, or ethnic interests; and, in the realm of the intellectual and social, the existence of international cultural elites.

The cultural components of postcolonialism are, in some ways, the hardest to recognize and certainly the hardest to change, not least in regard to the architectural and spatial attributes of urban form. To understand this means paying some attention to the process of colonialism itself.

In what had previously been pre-industrial societies around the world, the forms of colonial urbanism introduced in the nineteenth and the twentieth centuries were generally, in the later phase at least, forms of western industrial capitalism, varying according to the cultures from which they came and the colonial circumstances of their introduction. Implanted in Asia or Africa next to the indigenous city (if there was one), they created a dichotomous structure that is constantly misinterpreted, subjected to a category confusion and labelled according to temporal or geographical criteria, i.e., 'modern' (in relation to the 'traditional') city, or 'Western' (in relation to the 'non-Western') settlement, and not according to the mode of production (or system of socio-economic organization and political control) of which they were the outcome (i.e., early mercantile colonialism or late industrial colonialism). Consequently, the terms 'modern' and 'traditional' have been overinvested with political, cultural and psychological meaning.

Moreover, forms of social and cultural practice developed within the largely capitalist forms of urban development in the metropolitan societies (surveying, planning, architecture and building, as varied as these may have been between different cultures) and the knowledge on which they were based were established in the colonial societies, gradually replacing the practices and knowledges which once belonged to the indigenous society. Professional practice in architecture and planning, universally referred to as 'Western' or 'International,' has its origins and assumptions in the market institutions of capitalism.

The analysis of cultural production under the conditions of postcoloniality has recently become the focus of attention, especially in regard to literature. Though not addressing the fundamental questions about cultural production under industrial capitalism, the aim of recent critical writing is to displace Eurocentric canons of judgement, reiterating the importance of orality (compare, for example, the emphasis in architectural discourse on 'traditional building methods'), and to contest notions of 'universal' criteria in writing (this also highlights the naïvety of what has gone down in architectural history as the 'International Style'). The emphasis in research on postcolonial cultures is on distinguishing the political, economic and cultural features which act on different forms of cultural production (music, writing, dance, but also architecture and spatial organization) within the overall imperial framework. In emphasizing the decentering of the dominant cultural perspective, some postcolonial theorists also share and utilize feminist perspectives, questioning not only the predominance of capitalist, Western, Eurocentric assumptions but also the male episteme (Ashcroft, Griffiths and Tiffin, 1989; Mani, 1989; Spivak, 1990). In the urban sphere, new attention to questions of gender in the organization of colonial (and subsequently, post-colonial) space highlights the essential masculinity of colonial urban spaces (with the racecourse in the place of the CBD or the military cantonment appropriating significant areas of the colonial capital) (Calloway, 1989).

Postimperialism

The postimperial paradigm addresses the urban situation of once metropolitan centers (the imperial capital, such as Paris, Tokyo, London, Amsterdam, Washington, Berlin, Vienna, Madrid) as well as other urban places which, in terms of economic, political, social and cultural function, played a critical role in the imperial space economy and culture (King, 1990b). This may be in terms of the industrial processing of colonially-derived cash crops (cocoa, tobacco, cotton, wool, timber, cereals, etc.) in particular portcities or regions; government, financial and administrative functions (colonial control, banking, investment); communication and transportation functions (shipping, telecommunications); intellectual and educational functions and the general production of knowledge (universities, publishing).

In the environment of the post-imperial city, such functions are represented, for example, by the concentration of international banks, insurance houses and headquarters of multinational (once imperial) companies; by institutions for the study of tropical medicine and hygiene or 'Third World' development; by museums and centers of higher learning focussing on the study of one-time colonial cultures or on the institutional development of multiculturalism (not least in relation to the new demographics of post-imperialism which bring increasing numbers of once colonial populations to work, study and reside in what had previously been represented as the center of their universe). It is in, and on the basis of experience of, the post-imperial city and society that new paradigms of national cultures, multiculturalism and globality are constructed (and contested), frequently by postcolonial intellectuals and critics.

Postmodernism

The notion of the 'postmodern' is premised on the assumption of the 'modern,' which, in contemporary usage, is grounded in a very distinctive industrial and monopoly phase in the development of the capitalist world-economy which took place in Europe and the USA. If 'modernity' can be defined by reference to the cultural practices of specific elites (an arguable proposition), then, in Harvey's argument (1989), which is probably the clearest exposition of the postmodern thesis, it was a cultural manifestation of twentieth-century capitalism which, though premised on relations within the world-system (in colonialism), was nonetheless restricted in its manifestations to Europe and North America.

Yet 'modernity' in these continents was never understood in this way (see, for example, the representative text *Modernism*, 1890–1930 by Bradbury and McFarlane, 1976). It was invariably defined only in relation to Europe and the USA and not within the world-system as a whole. It follows, therefore, that 'postmodernity' operates only within the same geographical restrictions, whether in terms of the Eurocentric intellectual sources on which it draws, the phenomena it purports to explain or the areas of the world to which it relates. J. F. Lyotard himself states that 'the postmodern condition' (1984) is only a symptom of 'the most highly developed societies' and for Frederick Jameson, the other high priest of the postmodern, it is 'almost a synonymous term for American culture.'

This is not, therefore, a conceptual paradigm which can be used to discuss global-level processes, even though, *as* an intellectual paradigm and a particular design practice, it has, in fact, become part of the globalization process (see below) and, curiously, has assumed the mantle of privileged critic of the Western episteme (Connor, 1990).

These five theorizations do not, of course, exhaust the possibility of attempts to construct conceptualizations of the globe, globality or the world. Luard (1990), for example, writes of the 'internationalization of society,' and recent theorizations of development have invoked the much earlier theories of global convergence put forward in the 1950s by Teilhard de Chardin (Terhal, 1987). They do, however, provide alternative conceptual spaces within which to identify some of the universal as well as particular characteristics of individual world cities.

Drawing selectively on these distinct but also interrelated frameworks, therefore, we can now consider recent developments in three major cities, London, New York and Delhi, transformations of an economic, demographic, social, cultural, as well as physical and spatial nature.

London

London is best understood as a post-imperial city whose principal function in the world-system is as banking and finance center. Yet where, as recently as 1980, this was largely contained within the boundaries of the City of London, developments in the 1980s (technological globalization and its corollary of deregulation, the coming impact of the single European Market in 1992 and the political transformation of Eastern Europe) have extended the influence of the City into much more of London's economy as well as its urban space.

London's financial role in the world-economy goes back at least 200 years and is closely tied to Europe's economic and political role as well as Britain's imperial history and trade. The representation of foreign banks in London (from the early 1800s) was an outcome of imperial expansion, though from 1806 European and other banks moved in, followed at the end of the century by American and Japanese banking houses, gradually building up to number over 100 by the midtwentieth century (King, 1990b, from which the following paragraphs are drawn).

The huge expansion in London as a financial center, however, has occurred from the 1960s with the growth of the Eurocurrency market. This brought first American and then Japanese institutions into a less regulated financial market than existed in their home countries, financing trade and funding long-term capital. Between 1960 and 1980, the number of foreign banks or representative offices soared fourfold, from about 100 to well over 400, and by the late 1980s there were some 450, employing over 50,000 people. In 1989, London accounted for 20 percent of total world international banking (twice that of the U.S.) though the competitive nature of this activity is seen when this figure is compared to the 27 percent which it had in 1980. During these nine years, New York's share fell from 14 percent to 10 percent, while that of Tokyo rose from 5 percent to 21 percent (Diamond, 1991, p. 88; Green and Hoggart, 1991, p. 222).

Despite this buildup in international banking, foreign currency dealing (of which London had 43 percent of the world share at the end of the 1980s; Diamond, 1991) and other business, because of the historic monopoly exercised by the London Stock Exchange, London's share of global securities dealing in the early 1980s was still only one fifth that of Tokyo and a mere one twelfth of that of New York. The Conservative Government's decision to deregulate the Stock Market was to have immense influence, not only on the massive growth of the financial services industries and related employment in legal, accounting and insurance services, but, as Darrel Crilley shows in Chapter 6, on real estate, the construction industry and architectural design.

Along with the increase in foreign banks, the number of foreign security houses dealing in London also grew, from 10 in 1960 to 76 in 1980, at first mainly from the USA and Canada and subsequently from Japan. With the 'Big Bang' of 1986, this number almost doubled, to 120, in 1987 and, with the expansion in the number of banks, insurance and business services, a huge construction boom developed in London in the mid-1980s, particularly in the City, with the critical decision in 1985 to develop an additional 10 million square feet of office space at Canary Wharf in the Docklands (see Chapter 6). This complex now provides for the European headquarters of American multinational banks, Morgan Stanley and American Express, as well as the New York architectural firm Skidmore, Owing and Merrill. In the City, the new 'banking factories' requiring 500,000 square feet in one slab to provide uninterrupted dealing floors (for conglomerates such as Citicorp, Express, Nomura) was to totally change the face of specific areas in the City.

Yet the crash of October 1987 and subsequent recession have likewise had an impact on the scale and nature of London's space and built environment as well as its economy and occupational structure in general. Many of the foreign banking groups that plunged into market-making in 1986 lost money and left. By 1990, the number of security houses had shrunk to about 100, with consequent negative effects on the already sluggish property market. Moreover, in the run up to 1992, London, traditionally relatively free from regulation, was to face a growing body of regulations from the European Community; with increasing competition from trading markets in Paris, Amsterdam, Switzerland, Stockholm, Copenhagen and Frankfurt, the City has become increasingly conscious of continental competition. In 1989, *The Banker* (November) pointed out that the largest contingent of foreign banks in London (235 out of some 450 branches) were from Europe, a substantial change from the 1960s when they were primarily American or, earlier in the 1980s, Japanese. Between 1985 and 1989, these European banks raised their share of foreign currency dealing from 27.7 percent to almost 33 percent. London-based European banks have become increasingly important as lenders to British manufacturing and construction concerns.

These developments have gained added importance since the barriers to Eastern Europe and the Soviet Union began to crumble in 1989 and capitalist institutions marched eastward. Stock markets were opened in Leningrad in March 1991 and in Warsaw shortly afterward. After some initial skirmishes with Amsterdam in spring 1991, the London banking lobby won the battle to locate the European Bank for Reconstruction and Development in London (larger in terms of number of directors than the International Bank of Reconstruction and Development).

Competing threats to London's role as preeminent European financial center, particularly from Paris and Frankfurt (and Berlin, named as the new capital of a reunited Germany in Spring 1991) are not to be underestimated. In the Europe of the twenty-first century, London will be on the Western periphery. In 1991, one in ten Frankfurt jobs were in financial services and Frankfurt was bidding to host the European Central Bank (being established to preside over the single European currency). Two thirds of the banks there are foreign and, unlike London, Frankfurt's financial function is not distracted by political change. With the economic strength of Germany behind it (temporarily dampened by demands from the East), Frankfurt is a serious contender as a European financial capital; it is the seat of the German central Bank, the German futures and options exchange and the home of 410 financial institutions, including 270 foreign banks. However, unlike London, banks in Germany in 1991 were still disadvantaged internationally in terms of the proportion of their lending portfolio required to be deposited with the central Bank: hence, deutschmark business was still done abroad. Nonetheless, the aspirations of Frankfurt to be the New York of Europe were represented by the new, 260-meter Postmodern Messe Tower, built to promote Frankfurt's case in the new Europe of the 2lst century (Milner, 1991).

Irrespective of these developments, Japanese banks have continued to move into London, described by *The Banker* in November 1990 as the Japanese capital of Europe. Between 1984 and 1989, Japanese banking houses and representative offices increased from 23 to 85, expanding rapidly into property lending, construction, venture capital projects, and the investment space created by the privatization of Britain's previously nationalized industries. From 1975 to 1988, Japanese banks' share of international lending out of London almost tripled, from 13 to 36 percent (Diamond, 1991, p. 88). In these developments, the Japanese construction company Kumagai Gumi, whose meteoric rise rests on its ability to translate international finance capital into the built environment (Rimmer, 1990), has been a significant player.

The uncertainty brought by the world recession, the Gulf War and the collapse of stock markets globally in the early 1990s meant that many of the specialized buildings of the mid- and late-1980s designed for use for dealing have been remodelled as cellular office space. The potential growth in legislation emanating from Brussels as 1992 approached was seen to be generating demand for legal services and hence for legal offices; according to *The Banker* (1990), there is 'a global reliance on London lawyers for a range of financial instruments and products written under English law.'

In the early 1980s, developers and clients still hesitated about locations outside the prestigious historic boundaries of the traditional City. However, the demand for much larger quantities of office space at competitive prices broke the magic thread and after 1986 led to huge developments in the West End of London and on the hitherto unacceptable South Bank of the Thames (marketed in the late 1980s as 'the transpontine area'). With Canary Wharf, in Docklands, the office market has expanded unprecedentedly, even though, after Tokyo, London office rents are the highest of all world cities and London is one of the most expensive cities after Tokyo. By the end of the 1980s, with major plans for the development of the Kings Cross area, it was widely reported that the historic division between the City and central London was disappearing; the prospect of a single market in commercial property, from Aldgate in the east to Victoria in the West, with all that this implied, was round the corner (Diamond, 1991). Demand came not only from banks, business services and multinational companies, but also from expanding cohorts of accountants and lawyers whose practices have been increasingly transnationalized (Dezelay, 1990).

Despite the sluggish nature of demand and the prospect of between six to nine million square feet of office space coming onstream in 1991, Japanese and, increasingly, Scandinavian companies are putting investment into London, with an emphasis on what have been termed 'trophy buildings'—design-intensive statements, the symbolic capital of global capitalism at the end of the twentieth century.

In addition to its global banking function, London is also a pre-eminent command and control center, localizing the headquarters of most of the 190 UK-owned firms in the largest 500 European companies (Hamilton, 1986, p. 54). As Hamilton points out, 'London also outshines New York where, since 1970, the number of headquarters of Fortune 500 industrial firms fell sharply from 118 to 48. Even if the Fortune 500 service companies are included, only 110 of the 1000 largest US firms are headquartered in New York. That does not compare with London which, in 1988/9, had 41 percent of the Times 1000 firms in the UK, though this represented a decline from 53 percent in 1971/2.' In that interval, most headquarters decentralized to other parts of South East England, which, including London, contains 60 percent of the total (Hamilton, 1986, p. 54).

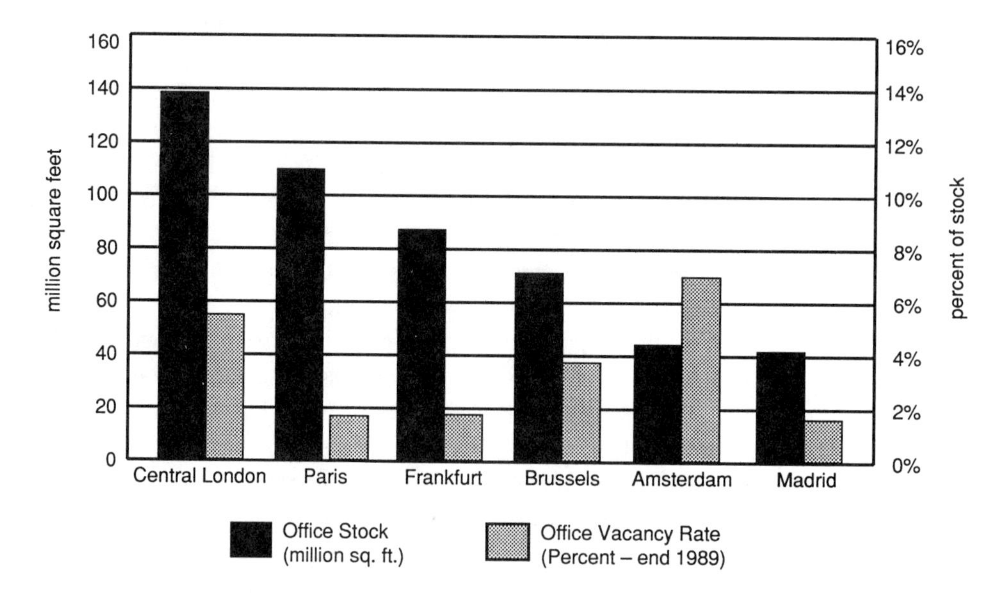

Fig. 4.1. European office and property markets. *Source*: London Planning Advisory Committee.

London now competes in two interlinked global markets. In the financial and business sector, its rivals are New York and Tokyo, both of which offer a much greater stock of office space (Manhattan has eight times the space of the City of London: Duffy and Henney, 1989, p. 22). In Europe, if only space is considered, its closest rivals are Paris and Frankfurt, yet with 13 million square meters (140 square feet), Central London has substantially more than either of these and only a little less than Brussels, Amsterdam and Madrid together (Fig. 4.1). However, as Diamond (1991) points out, new informational technology demands have made much of this out-of-date or obsolescent.

The outcome of these developments and the continuing flight of national 'non-profit' organizations from London (driven out by global competition, escalating prices of real estate, especially affordable housing, the absence of any strategic governing body for London, and rapidly increasing economic and social polarization with the growing armies of homeless) are all issues subject to political values and decisions. Yet they are decisions made within larger historical conditions and these are not of any particular government's choosing.

New York

London's role in the internationalization of capital has a significant impact on its built environment. Can it be argued that New York occupies a particular space and performs a particular role in the globalization of culture, which also influences its built environment? And does its built environment, whether as image or reality, contribute in some way to the process? These questions are addressed in the following section.

In the everyday sense, 'culture' is generally taken to refer to either or both of two phenomena: the arts, literature, music, the media, etc., and, in anthropology and the social sciences more generally, ways of life and the values and beliefs which inform them. More sophisticated interpretations emphasize the way culture, as a code of signs imbued with meaning, is socially produced and represented, providing social subjects with different cultural identities. Insofar as culture taken as 'the arts, music, literature' draws on, (as well as participates in) the social construction of culture as a way of life, as a system of values and beliefs which, in turn, affect culture as a creative, representative practice, we can bridge the gap between these two different meanings (King, 1991a). In what follows I shall draw on each of these differentiated interpretations. The issue to be addressed is how far and in what ways certain world cities, and New York in particular, exercise cultural hegemony on a global scale.

Gramsci's original notion of hegemony implied the domination of one class over another by both political and ideological means. The chief instrument of the coercive force was the state, winning popular consent by ideological domination through the institutions of civil society, the church, family, schools and so on (Lears, 1985).

As has been stated elsewhere (King, 1992),[2] there are three realms where we might consider the city to have a hegemonic cultural role: the first considers the city as a distinct, and influential, social and cultural formation; the second, as the site for the accumulation of cultural capital (Bourdieu, 1977); the third, as built environment, symbolic capital and space.

In their classic article on 'the cultural role of cities,' Redfield and Singer (1954) distinguished two basic city categories: the orthogenetic (the city of the moral order, of culture carried forward) and the heterogenetic (the city of the technical order where 'local cultures are disintegrated and new integrations of mind and society are developed').

As heterogenetic formations, world cities manifest the highest degree of cultural plurality in terms of the significant proportions of different racial, ethnic and religious groups present in them. Yet how these different cultures develop, how cultural plurality is constructed and institutionalized, either theoretically, in discourse, or practically, in social and educational policies, for example, obviously differs according to each city and circumstances. The historic difference between, say, South Africa's Cape Town and Brazil's Rio de Janeiro is sufficient illustration of this. The way in which cultures are constructed clearly depends on the political, economic, social, racial, as well as physical and spatial conditions prevailing at any one time, in any one city or state.

In this respect, the range of ethnic and cultural diversity of the New York City population is probably unrivalled. According to the Director of City Planning, New York City is 'by far the most ethnically and racially diverse city in the world' (*New York Herald Tribune*, 23 March 1991). The largest single group (as constructed according to current 'official,' culturally-specific systems of classification) is 'non-Hispanic whites' (43.2 percent); other principal groups including blacks (25.2 percent), Hispanics (24.4 percent) and Asians (6.9 percent). Of the foreign born adults in 1988, almost half were from the Caribbean and Central America, with other significant proportions from Europe (21 percent), South America (15 percent), Asia (12 percent) and Africa (4 percent). Of these groups, the largest numbers in 1980 (with populations of more than 80,000) included migrants from Italy, the Dominican Republic, Jamaica, the USSR and China; with between 40,000 and 80,000: Poland, Germany, Haiti, Cuba, Ireland, Greece, Colombia, Trinidad and Tobago and Equador; and with about 20,000: Guyana, Austria, England, Hungary, Yugoslavia, India, Philippines, Panama, Korea, Barbados and Rumania. This is within a total population of 7,322,564 (Bayer and Perlman,1990).

A population with such a wide range of cultural, linguistic, ethnic and racial origins,

[2] This section draws on King (1992).

with the proportion of foreign born expected to reach a projected 53 percent of the total city population by the year 2000 (compared to 23.6 percent in 1980; Bayer and Perlman, 1990), confirms New York as containing a truly international representation of cultures. This is particularly so in comparison to London, which, with a relatively high proportion of inhabitants of Asian, African and Caribbean origin, suggests a form of multiculturalism which is more post-colonial or postimperial than international. (In 1981, the proportion of London's population living in households headed by a person born overseas was about one quarter.)

The conditions under which different cultural forms, ethnic identities and representations of multiculturality are being constructed, however, are not only peculiar to the USA (with its heightened sense of the market, history of slavery, etc.) but also to New York City, including its built environment and historic ethnic districts. The relatively recent shifts, for example, from social and educational policies promoting the 'melting pot' to others celebrating ethnic and racial 'diversity' are adequate illustration of this. In this context, few recent instances of commodification can rival, in the capitalist world in general and the USA in particular, the rapidly growing market in ethnic difference. (Others will have noted the increase in anti-Semitism which has accompanied the demise of authoritarian regimes in Eastern Europe.)

The issue which this raises for an increasingly multicultural, polyethnic urban world is the extent to which the multiethnic cultures, forms of knowledge and social or educational policies being constructed in New York are being transplanted elsewhere, particularly to other world cities in the English-speaking world. In a larger sense, specific cities (e.g., New York, London, Paris, Rio—all, incidentally, with colonial or imperial pasts) become the privileged sites for the production of transnational cultures, privileged because they operate with 'world' (i.e., ex-colonial) languages.

The second sphere in which we might think of cultural hegemony and world cities concerns the city as site for the accumulation of cultural capital, the massively expanding realm of the global cultural economy, the ideas, images, and signs which form an ever-increasing proportion of post-industrial economies and cultures: the information, advertising and communication industries, the world of the fine and applied arts, film, video, TV, disc, tape, cable, publishing—a sector which, in the economies of world cities, is increasingly oriented to foreign markets. In the mid-1980s, the foreign business of the top ten advertising agencies accounted for half their revenue.

Here, I refer not only to culture in the sense of arts and music, but also educational, economic, management and design culture: the world of financial instruments, forms of knowledge and education. As cities and higher-educational institutions position themselves globally, knowledge production moves more consciously into the world market, transnationalizing the curriculum; architecture, planning and design services are increasingly marketed on a global scale.

In New York, as elsewhere (GLC, 1985), the cultural industries are now seen as critical to the urban economy, supported as they are by the burgeoning financial services sector as well as tourism (Port Authority of New York, 1983). At a global level, 'packaged' culture overwhelmingly comes from the industrialized countries, which, though making up only 36 percent of the world's population, produce 83 percent of all book titles, the USA (with the UK and the Soviet Union) consistently among the top three publishers. In the global promotion and advertising of consumer goods, the US is seen as the model, the rest of the world rapidly emulating its practices. The vast part of this industry is generated from New York. Mowlana (1986), from whom these data are taken, concludes, 'The ability to continue the means of production and international distribution of cultural products then is the key to larger markets and greater productivity in an international system that has eschewed "gunboat" coercion, to some extent, in favor of the utilization of cultural industries as persuaders' (p. 2).

Where cities on the periphery are still actively engaged in nation-building, global cities in the core are actively engaged in world-building. In the more conventional understanding of culture (the arts, music, theater, literature, dance) New York has frequently represented itself as the hegemonic cultural capital, not least in the collection of essays edited by Leonard Wallock: *New York. Culture Capital of the World, 1940–1965* (1988). Leaving aside the unilateral nature of such claims or the privileged definitions of culture on which they rest, few would venture such a statement today. The immensely differentiated pattern of cultural activity in the contemporary world, as Robert Hughes (1990) points out, has made the very idea of the single imperial center obsolete even though New York does remain a market center, 'an immense bourse on which every kind of art (is) traded for escalating prices.'

Similar comments may be made in regard to New York's role as global architectural paradigm, at least outside the USA. If international capitalism gives rise to a particular culture which is represented in space and architectural form and if, as part of a global system of production, its elements may be found all over the world, its representative center for many years was certainly New York, the host of 350 international banks as well as many multinational headquarters. Yet as indicated above, many corporate headquarters have left New York since 1970. And where the city's distinctive skyline, already established by the 1920s, became the symbol of corporate capitalism as well as a model for 'modernity' worldwide by the middle of the twentieth century (not least through the global projection of its image through film and photograph), there are now competing alternatives of development. Nations and cities have become more conscious of their own histories and identities, questioning the previously unquestioned assumptions of a Western 'modernity.' These developments are themselves an outcome of globalization, as also are the batteries of comparative data on world cities and other urban phenomena which have been generated since midcentury. Images of what might constitute 'the modern' have been complemented by concerns about 'livability': adequate housing and services, clean air, safety, employment and the absence of noise. Here, in a recent survey, New York ranked 27th out of the world's 100 largest (2 million plus) cities (Population Crisis Committee, 1990). Arguments supporting New York's global cultural hegemony, therefore, seem partial, at most.

Delhi

The world city of India is Bombay rather than Delhi. Yet some reference to both is needed to demonstrate why different, though overlapping global frameworks are needed to understand recent transformations in the built environment of India's capital city.

Even before the mid-nineteenth century, Bombay had grown to be the economic and financial center of India, the principal link to London as the banking capital of Empire. In the late 1980s, it hosted 19 of the 21 main offices of foreign banks in India. It has also the largest number of branch offices (45), though only one more than Calcutta, with Delhi (19) in third place. In 1985, a new, 22-story Stock Market building was opened in Bombay (King, 1990b).

Yet Delhi, as the national capital, has also been the site of immense investment in the 1980s, the result of economic growth which has contributed to the creation of a substantial middle class market, variously estimated at between 60 and 100 million people in a total population approaching 1,000 million.

Between 1970 and 1990, the built environment of what was once an archetypical colonial city, previously the capital of the British Raj, has gradually been transformed into that of a South Asian capitalist city. Foreign banks, four in 1970, are now numbered in double figures; they are some of the principal tenants of new high-rise blocks which have transformed the city center. In the first half of the twentieth century, the classical colonnades of Connaught Place ringed the retailing center; as Delhi moves into the twenty-first century, these are making way for the high-rise towers of indigenous enterprise (Fig. 4.2). Of the foreign banks, the most spectacular is the recently-completed Citibank, which, though a joint venture with majority Indian ownership and linked to the Indian Life Insurance Building, has brought 1980s glass-walled modernism alongside the rundown columns of Connaught Place (Fig. 4.3). Further from the center, Western- (or internationally?) oriented architects have produced the first postmodern façade.

Fig. 4.2. The neo-classical colonnades of Connaught Place give way to high-rise Modernism.

The frames for understanding these changes are best provided by a combination of postcolonialism and globalization. The architectural mutation of New Delhi is selective. Where extensive areas of the colonial space of Connaught Circus are demolished or neglected, one quadrant has been restored to its earlier splendor (Fig. 4.4), the owner (or tenant), British Airways, posting howkidars (guards) to patrol the sidewalk, moving itinerant 'panwallahs' and beggars to territories on other sites.

Fig. 4.3. The Citibank building, Delhi.

To the south, an entirely new city has sprung up since the mid-1980s, including a string of 'farm-houses,' so called from their location in extensive walled and gated estates which form a 'green belt' cordon round the southern extremities of New Delhi. Here, a number of rapidly growing housing projects have been constructed for middle and upper class occupancy. The architecture is a hybridized mix from Bombay, South Kensington and the Gulf states, from which some of the capital is likely to have been repatriated. With tens of thousands of its nationals working and earning (as 'non-resident Indians') in different countries abroad, the styles of contemporary architecture in Delhi are certainly global.

Fig. 4.4. A restored quadrant of Connaught Circus.

To ask, therefore, almost fifty years after Independence, where is New Delhi's (and India's) cultural identity (at least as this is represented in its architecture and urban design) is to realize that it is in the totality, and variety, of all that actually exists. It is, to make use of Featherstone's suggestion (1990), one of a number of global cultures, yet one that is very locally expressed.

References

Abou-El-Haj, B. 1991 Languages and Models for Cultural Exchange, in Anthony D. King, ed., *Culture, Globalization and the World-System, Contemporary Conditions for the Representation of Identity*. Binghamton NY, Department of Art and Art History, SUNY Binghamton and London: Macmillan, 139–144.

American Journal of Sociology 1975 Review symposium on Immanuel Wallerstein, *The Modern World-System*, 82, 1075–1102.

Ashcroft, B., G. Griffiths, and **H. Tiffin** 1989 *The Empire Writes Back, Theory and Practice in Postcolonial Literature*. London: Routledge.

Bayer, D. and **J. Perlman** 1990 Here is New York — 1990. Paper presented at the Conference on Megacities of the Americas. State University of New York at Albany, April 5–6, 1990.

Berry, B. L. 1990 Comparative Geography of the Global Economy: Cultures, Corpora tions and the Nation State, *Economic Geography*, 65, 1–18.

Bourdieu, P. 1968 Outline of a Theory of Art Perception, *International Social Science Journal*, 2, 589–612.

Bourdieu, P. 1977 *Outline of a Theory of Practice*, trans. Richard Nice. Cambridge: Cambridge University Press.

Boyne, R. 1990 Culture and the World-System, in Featherstone, M., ed. *Global Culture. Nationalism, Globalization and Modernity*. Newbury Park, CA: Sage, 57–62.

Bradbury, M. and **J. McFarlane** 1976 *Modernism. 1890–1930*. Harmondsworth, Penguin.

Braudel, F. 1984 *The Perspective of the World*. London: Fontana.

Calloway, H. 1989 *Gender, Culture and Empire*. London: Macmillan.

Clifford, J. 1989 Notes on Theory and Travel, *Inscriptions*, 5, 177–88.

Cohen, R. B. 1981 The New International Division of Labor, Multinational Corporations and Urban Hierarchy, in M. Dear and A.J. Scott, eds., *Urbanization and Urban Planning in Capitalist Society*. London: Methuen, 287–315.

Connor, S. 1990 *Postmodern Culture*. Oxford: Blackwell.

Cooke, P. ed. 1989 *Localities*. London: Hutchinson.

Dezalay, Y. 1990 The Big Bang and the Law: The Internationalization and Restructuring of the Legal Field, in Featherstone, M. (ed.), *Global Culture. Nationalism, Globalization and Modernity*. Newbury Park, CA: Sage, 279–94.

Diamond, D. 1991 The city, the 'Big Bang' and Office Development, in Hoggart and Green, *London. A New Metropolitan Geography*, 79–94.

Duffy, F. and **A. Henney** 1989 *The Changing City*. London: Bulstrode Press.

Featherstone, M. 1990 Global Culture: An Introduction, in M. Featherstone (ed.), *Global Culture, Nationalism, Globalization and Modernity*. Newbury Park, CA: Sage, 1-14.

Frank, A. G. 1991 Transitional ideological modes: feudalism, capitalism, socialism, *Critique of Anthropology*, 11, 171–88.

Friedmann, J. 1986 The world city hypothesis, *Development and Change*, 17, 69–83.

Friedmann, J. and Wolff, G. 1982 World City Formation: An Agenda for Research and Action, *International Journal of Urban and Regional Research*, 6, 309-44.

Geddes, P. 1949 *Cities in Evolution*, London: Williams and Norgate (first edition, 1915).

Gottman, J. 1989 What are Cities Becoming Centers Of? Sorting out the Possibilities, in Knight, Richard and Gary Gappert, eds. *Cities in a Global Society*. Newbury Park, London and Delhi: Sage, 58–67.

Greater London Council 1985 *The London Industrial Strategy*. London: GLC (Ch. 6. The Cultural Industries).

Green, D. R. and **K. Hoggart** 1991 London. An uncertain future, in K. Hoggart and D. R. Green (eds.), *London. A New Metropolitan Geography*. London: Arnold, 220–32.

Hall, P. 1984 *The World Cities*. London: Weidenfeld and Nicolson.

Hamilton, A. 1986 *The Financial Revolution. The Big Bang World Wide*. Harmondsworth: Penguin.

Harvey, D. 1989 *The Condition of Postmodernity*. Oxford: Blackwell.

Henderson, J. and **M. Castells** 1987 *Global Restructuring and Territorial Development*. London, Newbury Park and Delhi: Sage.

Hoggart, K. and **D. R. Green**, 1991 *London. A New Metropolitan Geography*. London: Arnold.

Hughes, R. 1990 The Decline of the City of Mahogony, *The New Republic*, June 25, 1990, 27–38.

Jones, E. 1990 Review of A.D. King, *Global Cities, Journal of Historical Geography*, 16, 454–5.

King, A. D. 1990a *Urbanism, Colonialism and the World-Economy*, London and New York: Routledge.

King, A. D. 1990b *Global Cities, Postimperialism and the Internationalization of London*. London and New York: Routledge.

King, A. D. (ed.) 1991a *Culture, Globalization and the World-System, Contemporary conditions for the Representation of Identity*. Binghamton NY: Department of Art and Art History, SUNY Binghamton.

King, A. D. 1991b The global, the urban and the world, in King, A.D. (1991a), *Culture, Globalization and the World-System, Contemporary conditions for the Representation of Identity.* Binghamton NY, Department of Art and Art History, SUNY Binghamton and London: Macmillan, 149–54.

King, A. D. 1992 Culture Hegemony and Capital Cities, in National Capital Commission, *Capital Cities*. Ottawa: University of Ottawa Press.

Knight, R. V. 1989 The emergent global society, in Knight, R. V. and G. Gappert (1989), *Cities in a Global Society.* Newbury Park, London and Delhi: Sage, 24–43.

Knight, R. V. and **G. Gappert** 1989 *Cities in a Global Society.* Newbury Park, London and Delhi: Sage.

Lears, T. J. 1985 The concept of cultural hegemony: problems and possibilities, *American Historical Review*, 90, 567–93.

London Planning Advisory Committee 1990 *Strategic Trends and Policy, 1990 Annual Review*. 1990 Annual Review. Romford: LPAC.

Luard, E. 1990 *The Globalization of Politics*. London: Macmillan.

Lyotard, J.F. 1984 *The Postmodern Condition*. Manchester: Manchester University Press.

Mani, L. 1989 Multiple Mediations: Feminist Scholarship in the Age of Multinational Reception, *Inscriptions*, 5, 1–27.

Milner, M. 1991 The Challenge to the City, *The Guardian*. April 6, 1991, 26.

Mowlana, H. 1986 *Global Information and World Communication*, New York and London: Longman.

New York Herald Tribune 1991 New York's Minorities Have Become Its Majorities. 23 March 1991.

Population Crisis Committee 1990 *Cities, Life in the World's 100 Largest Metropolitan Areas*, Population Crisis Committee, Washington.

Port Authority of New York 1983 *The Arts as Industry, Their Economic Importance to the New York-New Jersey Regional Area.* New York: Regional Assistance Center, Inc.

Redfield, R. and **M. Singer** 1954 The cultural role of cities, *Economic Development and Cultural Change*, 3, 53–73.

Rimmer, P. J. 1990 The Internationalization of the Japanese Construction Industry: The Rise and Rise of Kumagai Gumi, *Environment and Planning A*, 22, 345–68.

Robertson, R. 1990 Mapping the Global Condition: Globalization as the Central Concept, in Featherstone, M., (ed.), *Global Culture. Nationalism, Globalization and Modernity.* Newbury Park, CA: Sage, 15–30.

Robertson, R. 1987a Globalization theory and Civilizational Analysis, *Comparative Civilizational Review*, 17, 20–30.

Robertson, R. 1987b Globalization and Societal Modernization, *Sociological Analysis*, 47 (S), 35–43.

Sassen, S. 1991a Cities in a World Economy. New York, London, Tokyo. Paper presented to the Joint ACSP and AESOP International Congress, Oxford, July, 1991.

Sassen, S. 1991b *The Global City, New York, London, Tokyo.* Princeton: Princeton University Press.

Sassen-Koob, S. 1986 New York City: Economic Restructuring and Immigration, *Development and Change*, 17, 85–119.

Smith, M. P. and J. R. Feagin, (eds.) 1987 *The Capitalist City*, Cambridge: Blackwell.

So, A. Y. 1990 *Social Change and Development, Modernization, Dependency and World System Theory.* Newbury Park, London and Delhi: Sage.

Spivak, G. 1990 *The Postcolonial Critic.* London: Routledge.

Terhal, P. H. 1987 *World Inequality and Evolutionary Convergence, A Confrontation of the Convergency Theory of Teilhard de Chardin with Dualistic Integration.* Delft: Eburon.

Wallerstein, I. 1974 *The Modern World-System, Capitalist Agriculture and the Origins of the European World-Economy in the Sixteenth Century.* New York, San Francisco, London: Academic Press.

Wallerstein, I. 1980 *The Modern World System II, Mercantilism and the Consolidation of the European World-Economy, 1600–1750.* New York, San Francisco, London: Academic Press.

Wallerstein, I. 1991a World System versus World-Systems: A Critique, *Critique of Anthropology*, 11, 189–94.

Wallerstein, I. 1991b The National and the Universal. Can There Be Such a Thing as World Culture? in King, A. (ed.), *Culture, Globalization and the World System.* Binghamton, NY: Department of Art and Art History, SUNY Binghamton, 91–106.

Wallock, L. (ed.) 1988a *New York, Culture Capital of the World, 1940–1965.* New York: Rizzoli.

Wallock, L. 1988b New York City. Capital of the Twentieth Century, in L. Wallock, ed., *New York, Culture Capital of the World, 1940-1965.* New York: Rizzoli, 17–52.

Wolff-Phillips, L. 1987 Why 'Third World'? Origin, Definition and Usage, *Third World Quarterly*, 9, 1311–27.

Young, R. 1990 *White Mythologies, Writing History and the West.* London and New York: Routledge.

Five

THE CITY OF ILLUSION: NEW YORK'S PUBLIC PLACES

M. Christine Boyer

Princeton University

Just walking through the streets of Manhattan is a challenge to any spectator's sensibilities—trash litters the gutters, smells emanate from sullied corners, the noise of honking cars and blaring music-boxes fills the air. There are as well the cardboard box shantytowns of the homeless, graffiti-covered walls, the outstretched hands and confronting pleas of panhandlers. But New York also displays unique landscapes of luxury and privilege. Since the city has just experienced its most prosperous decade since the Second World War, its signs of wealth and abundance are particularly apparent. Thus cheek by jowl with the city's visual displays of poverty lies the spectacular evidence that New York is still the financial capital of America—through its gateways flow national and international investment capital seeking new markets to exploit. And these investments in turn have generated in the last two decades swathes of new real estate development and redevelopment in large areas of the city such as Battery Park City, Times Square, the Upper West Side and South Street Seaport. In these districts the glimmer and glitter of wealth adorns new office skyscrapers, new hotels, new restaurants and theaters, new boutiques and retail shops, new luxury apartments and renovated townhouses.

Not to be denied, however, New York remains a city where the plight of the poor invades everyone's daily routine—it cannot be avoided or sheltered from sight. The city contains America's largest concentration of inner-city poor, revealing all the ravages of poverty be it homelessness, violent crimes, or abusive use of drugs. While many newspaper articles and critical reviews try to convince us that optimism in the city has vanished and that a deepening urban blight causes the city's quality of life to sink lower and lower in never ending decline, New York remains a city where dreadful nightmares and fantastic dreams invade each other's terrain. This is a city where there is no escape from the tangle of poverty and luxury, of crime and greed, of social justice and privatization. Every New Yorker is aware that the city's pathologies and its opportunities must be readdressed collectively, that each of us shares in the plight of the other.

The media image

Then why, we might ask, are sections of media so intent on detailing an upbeat image of the city? Why tap the city's nostalgic allure, its skyscraper scenography, and its special styles of life and cultural advantages in order to specify New York's unique distinctions of place? Let us look, for example, at an article written by Paul Goldberger, the architectural critic of the New York Times, entitled 'Why architecture can't transform cities.'[1] Let me briefly review Goldberger's position: with both the decline in the quality of urban life and indifference to the public realm rampant, architects and planners must adopt a 'new realism.' They must accept that there has always been a large gap between the rich and the poor in every great city. American cities, in particular, have always revealed a harsh and dirty appearance. And they must accept that over time these facts make the average citizen indifferent to the idea of the public realm and our shared commonality or plight in the city. Following Goldberger's argument, consequently, the homeless and vagrants have become simply 'part of the landscape,' just objects like lamp posts, that we basically ignore. We have no sense that architecture and planning are public services instilled with social responsibility for the well-being of others. The city's great liberal tradition producing the parks of Olmsted, the great public hospital system, City College, New York Public Library and Grand Central Station—this liberal tradition of the nineteenth century has been tarnished by an illiberal temper.

The reason for this tarnished tradition and the acceptance of this 'new realism,' so Goldberger argues, is the fiscal crisis of the mid-1970s: a time when the revenues collected fell far below the costs of paying for social services and providing for the needs of New York's poorer neighborhoods. In those years New York City was almost bankrupt. Since then the city has cut back on government spending until it no longer is the planner of the public realm, while the great majority of architects and planners now work directly for the private sector: building corporate skyscrapers, not public housing and public parks, designing consumer commodities for the well-to-do and appearing as publicity models for the advertising industry.

Goldberger hopes to convince us that architects and planners are scene-makers, not social reformers; they need no longer be concerned with the difference between public, semi-public or private space in the city; and they generally have lost the need or desire

[1] Paul Goldberger, 'Why Architecture Can't Transform Cities,' *The New York Times* (June 25, 1989): Section 2, pp. 1, 30.

for social accountability. From this point of view, architecture in the city is an autonomous artistic expression that has nothing to do with the social and economic plight of the poor. I would argue that this view is part of the reason why our collective view of the city has become fragmented into individualized pieces incompatible with the creation of a physical plan and why we have no map of the city linking together the poorer neighborhoods with the enclaves of the well-to-do. A plan or a map might draw us closer together and underscore our collective plight. It might help us point out the way towards a better future. But the city no longer plans for its physical development; it simply manipulates zoning bonuses and tax incentives that facilitate the building of huge real estate developments in ad hoc locations all over town. In return for these development gifts, the private developer must provide some of the amenities generally lacking in the public realm. Hence the 'new realism' is just this: we recognize, or so Goldberger claims, that the social programs of New York's more liberal tradition did not solve the city's social problems; but we also recognize without cynicism that the private sector does not really build in the public realm. 'We realize now, far more than in the last generation, that architecture must be evolutionary, not revolutionary, that it cannot make the world anew—We know there is no utopia.' Consequently, modest efforts bear sweeter fruits, and we now are resigned to work slowly bit by bit knowing that architecture counts, just not too much.[2]

Now, I have difficulties with this position, because it seeks to inscribe us as planners and architects and as readers or spectators of the new architecture of the city within a position of 'consensus,' where there is little we can do but accept the status quo, the gap that exists between rich and poor in a city where the quality of life declines with every new drug war, racial battle, and economic downturn. This kind of argument, furthermore, appropriates many discussions about architecture in the public sphere and normalizes these discourses, erases their critical distinctions and turns them to the purpose of consensus formation. It becomes a kind of public relations promotional for the new kinds of space being created throughout New York City, be they in Union Square, Times Square, South Street Seaport or Goldberger's favorite public space, Battery Park City.

For Goldberger, Battery Park City is an emblematic space—proof that we have not forgotten how to build a city in these postmodern times, how to develop a master plan that makes the public realm primary and architecture secondary. I will argue, using Battery Park City in particular, that this argument relies upon the art of dissimulation: making Battery Park City appear to be what it is not and appear not to be what it really is. This is what I mean by The City of Illusion: calling something public space when indeed it is not; focusing on the provision of luxury spaces within the center of the city

[2] Ibid., p. 30.

and ignoring most of the interstitial places; and (in the rush to create consensus around Battery Park City and other favored luxury projects in the city) not referring to the real problems of legitimation that haunt the latest stage of private multinational capitalism with its ability to shift capital around from country to country and region to region leaving swaths of uneven development, unemployment and bankruptcy in its wake.

Perhaps I should not blame a newspaper journalist for this artful dissimulation: for Noam Chomsky has pointed out in his book entitled *Necessary Illusions: Thought Control in Democratic Societies* that the mass media became a major target of reform during the years of Reagan's presidency. Bent into a non-adversarial position, the media now are supposed to appear supportive and properly enthusiastic for whatever programs the state sponsors. If the Reagan and the Bush presidencies have intentionally transferred resources to the wealthy, then the media have become 'vigilant guardians protecting [this] privilege from the threat of public understanding and participation.'[3]

The fragmented reality of the privatized city

Facts tell us a different story. The census has recently revealed what most of us have taken for granted—that during the 1980s, the top one-fifth of Americans' household wealth increased by 14 percent, while four-fifths remained the same.[4] The reality states that America is a divided nation of rich and poor, even though the media continues its style of artful dissimulation ignoring these trends. And while the media may focus on the lifestyles of the yuppies and guppies, gentrified neighborhoods and leisured pursuits, still an army of immigrants now compose over 30 percent of New York's population. Wall Street financiers, real estate brokers, lawyers and accountants may seem to make the city run; however it is by and large immigrant labor, most often underpaid or informally employed, who clean their offices at night, prepare their food in upscale restaurants, sew their clothes sold in trendy boutiques, man the taxi cabs that take them home late into the night or process their credit card accounts in back-office spaces. No matter how fragmented the city's spaces may be, New York is a city where the lives of the rich and the poor are undeniably linked.[5]

Behind this city fragmentation, and this denial of reality, lies another story of

[3] Noam Chomsky, *Necessary Illusions: Thought Control in Democratic Societies*. Boston: South End Press, 1989, p. 14.

[4] Robert Pear, 'Rich Got Richer in 80's; Others Held Even,' *New York Times* (Jan. 11, 1991): A1, A20.

[5] 'Despite Difficult Times, New York City Shows Unexpected Strengths,' *New York Times*, (Nov. 12, 1990): A1, B6, B7.

privatization. In the last two decades, transnational corporations facilitated by new information technologies have developed a global market for their goods and services and thus these super corporations have neither loyalty to a specific nation nor social accountability to any one locality. Many of these transnational corporations are major investors in the redevelopment of old American downtowns. Investment today, however, may bring abandonment tomorrow; hence these corporations need to build consensus that they are operating in the public interest, for the general welfare. If we can blame the fiscal crisis and the welfare state for bringing us to the brink of bankruptcy, then in this art of dissimulation private multinational capitalism may look like the friend of local authority. Such a good friend too, for the World Financial Center and new corporate headquarters glittering in the downtown centers of our cities are subsidized by local governments through tax write-offs, grants of special financial powers and abilities to override local land-use regulations, including a general disregard or lack of accountability to the public for the way these projects obtain and allocate their revenues. They have in many different ways privatized public decision-making.

And all of this public expenditure and deregulation for corporate benefit attempts to build consensus and attain legitimation both by seizing public events and public celebrations for corporate advantage and by privately providing a few greenhouse foyers within corporate high rises, a few 'living rooms of the city' as Philip Johnson refers to the atriums of buildings such as his AT&T or the IBM (both on Madison Avenue), and by offering a few historic theme parks and open places such as South Street Seaport to entertain workers during their leisure hours. Consequently, Goldberger claims that the 50 acres of new urban landscape at Battery Park City, the nearly two miles of esplanade that links a series of individually designed parks, public art and plazas, are demonstrations that New York City knows the importance of the 'public realm' and has been able to channel private money into public benefits. The World Financial Center Plaza and the Wintergarden, South Cove, North Park and South Gardens are examples of the new civic grandeur which corporate benefactors provide.

By adopting a language that speaks of the public realm, corporate developers try to make their projects appear to be promoters of local traditions and interests, not rivals with global alliances. Nevertheless, corporate expansion in the center of the city produces political and economic tensions in other parts of town. The growth of white-collar employment in the financial, insurance and real estate sectors means gentrification and the redevelopment of underutilized or devalued spatial territories of the city such as the old manufacturing edges, the air-rights over railroad yards, the seedy waterfronts and the honky-tonk parts of town. Turning these areas into luxury residential enclaves, entertainment zones or back-office spaces simultaneously extrudes and displaces the poor from the neighborhoods they formerly inhabited and pushes them into congested or peripheral parts of town. The manufacturing base of the city is restructured through global corporate growth, generating under- and unemployment in the traditional working class as jobs are destroyed.

In other words, this art of dissimulation mixes the public and private spheres together in great confusion. 'Public' in a democracy should refer to the entire populace, all groups, all neighborhoods, all regions of the country. Its access should be open and its construction untampered with. The public does not just mean the social welfare state or the government bureaucracy, nor does public space just mean space that is not private property. But in the last few years, the meanings of 'public' and 'private' have been confused: 'public' has almost become a bad word connoting unruly bureaucracies, corrupt officials, inefficient management, regulatory impositions, burdensome taxations while 'private' has turned into an exalted word: the freedom of the market, the freedom of choice and style of life that the market's commodities provide, even though we all know that the market is always a distorted provider delivering the most to those who already have a lot.[6]

Indeed in this war of positions, recruits are sought who support the dismantling and privatization of all of the social welfare programs of the last fifty years and accept that participation is restricted to private consumption. Consequently, the market economy is expected to provide everything: housing, health services, transportation, police protection, garbage collection, even public space. And the rhetoric surrounding Battery Park City, in particular, plays on this confusion of terms—creating a private preserve for the very wealthy that is transformed into 'a popular amenity' by allowing the people to stroll unimpeded along its corridors and spaces of power.[7] Since it has been the explicit policy of the federal government, in the last two decades, to gut the public sphere and promote the corporate sector, many sites and channels of public expression and creativity have been seized for private ends. Not surprisingly, we find that private corporations have invaded the cultural sphere, turning public spaces to market advantage. Once dependent on local sponsorship and community participation, now private corporations sponsor blockbuster museum exhibitions, underwrite newspaper advertising of gallery shows, control the open streets of shopping malls as private domains and orchestrate truncated public events and festivals. In New York, South Street Seaport, the Macy's Thanksgiving Parade, the new electric lighting of Times Square, special exhibitions at the Metropolitan Museum of Art or the galleries in the IBM building or Philip Morris, even Battery Park City, are really advertising events based on commercial, not cultural,

[6] Stuart Hall, *The Hard Road to Renewal.* New York: Verso Press, 1988.

[7] David W. Dunlap, 'Parking That Yacht' *The New York Times* (June 19, 1989): B3.

development.[8]

Now I will argue that the production of corporate-sponsored luxury spaces in the city must be linked to the increasing impoverishment of the poorer districts on the periphery. If private capital is able to subsidize and sponsor the public arts program in Battery Park City, for example, where sculpture gardens, horticultural displays and decorative lighting have been designed, then we must also be aware that since the late 1980s, public parks throughout the city are littered with broken glass, trash and abandoned cars, while the employment of maintenance workers has been reduced dramatically and the expenditure of tens of millions of dollars delayed.[9] Or to take another example that is reshaping the look of New York, employment figures for young urban professionals and white collar workers have spiralled significantly in the 1980s and developers have quickly responded by providing extensive residential and leisure spaces for these wealthy new recruits to city life. But simultaneously, there are many new immigrants from the Caribbean, Latin America, Asia and Africa (nearly 100,000 per year since the 1960s) who helped to sustain the economic boom of the 1980s. Most of these youthful new immigrants find employment in the under-paid and unregulated service and manufacturing industries such as restaurants, cleaning, taxi driving and garment industries—the very employment that enables the well-to-do to maintain their luxurious style of life. Unemployment among young urban blacks is at least 25 percent, contributing to the population that spills out of the ghetto to terrorize well-to-do enclaves of the city or to find financial reward along the Cocaine Trail.[10]

New York City in the last few decades has taken on an ugly appearance in many of its forgotten corners and neglected spaces. It suffers from an increasing number of single men and welfare families wandering its streets without a home. And while the City allows its supply of low-income rooming houses and single-room occupancy hotels to be renovated into luxury apartments, pushing more unfortunates onto the streets, it simultaneously locks the doors of its 'public' places to those who are forced to sleep over heating grates, in cardboard boxes, in parks and in subway trains. Even here the barriers of restrictions have risen: the Transit Authority has imposed new rules that prohibit begging and lying down on train seats, littering or creating unsanitary conditions or carrying

[8] Herbert, I. Schiller, *Culture, Inc. The Corporate Takeover of Public Expression.* New York: Oxford University Press, 1989.

[9] Andrew L. Yarrow, 'In New York's Parks, More Litter and Less Money,' *The New York Times* (Aug. 13, 1990): B1, B4.

[10] Richard Levin, 'Young Immigrant Wave Lifts New York Economy,' *The New York Times* (July 30, 1990): A1, B6.

out any unauthorized commercial activity and entertainment in spaces of public transit.[11]

Public space in urban design

Any contemporary reference to the 'public' is by nature a universalizing construct that assumes there is a collective whole, while in reality the public is fragmented into marginalized groups, many of whom have no voice, position or representation in the public sphere. Battery Park City, although touted to be a public space, stretches the concept. Impossible to enter from the street, it is accessible only by two elevated footbridges over West Street that tie it directly to the World Trade Center. Both bridges, if they can be found, lead the spectator into the platform lobby of the World Financial Center, which links the four office towers together one level above the ground. Still reminiscent of the futuristic megastructures of the 1960s, this arterial way is really an oversized and private interior hallway decorated in sensuous materials and bursting here and there into skylights and chandeliers, until arriving at the engineering feats of huge banks of elevators, stairways or escalators that cascade down to the street. There is, in addition, the glass-enclosed Winter Garden, squeezed in between two of the Center's towers, with its palm court exhibition and concert space, restaurants and shops offering New Yorkers one of the largest so-called 'public' atriums in the city. This awe-inspiring atrium has become the central focal point for all of Battery Park City and overlooks a 3.5-acre river plaza and The North Cove Yacht Harbor, where 26 yachts may rent a berth for the announced fee of $2.25 million apiece.[12]

Much more than displays of urban design are evident at Battery Park City. In the last few decades, for example, transnational corporations have become dependent on new systems of telecommunication in their daily national and international business activities. In order to insure their economic vitality, information pertinent to their business operations has been increasingly privatized and treated as commercial products. Simultaneously, the government's public information services have been largely eliminated and federal data bases privatized.[13] Herbert I. Schiller in *Culture Inc.* refers to this privatization of public information as the enclosure of sites and channels of public expression and creativity, making a lively comparison to the early nineteenth century enclosure of the English Common lands. But far more than information is being enclosed for commercial

[11] Sara Rimer, 'Pressed on the Homeless, Subways Impose Rules,' *The New York Times* (October 25, 1989), B5.

[12] Dunlap, op cit.

[13] Schiller, op cit. pp. 72, 83-84.

advantage and sales promotion. 'Corporate image and definition control' now cover an immense territory: from the information industry to museums to the streets and into our living spaces.[14] Systems of culture transmission under corporate sponsorship, without ties to specific localities or social identities and in an effort to legitimize their local operations, have invaded our cities and persuasively commercialized their public spaces. New festival marketplaces, museum atriums and shops, corporate foyers, public art, 'gardens' and interior arcades, in short public places like Battery Park City, are actually advertising sites linked to public relations campaigns in an attempt to humanize 'public-spirited' super-companies.

This new language of urban design follows formulas established by advertising and provides invented models of reality, seldom disguising their artifice. The city these spaces represent is filled with a magical and exciting allure, landscapes of pleasure intentionally separated from the city's more prosaic or threatening mean streets. Controlled by the rules and values of the market system, these places offer a diet of synthetic charm that undermines critical evaluation. Sumptuous architectural imagery, fictional information, entertainment and spectacle are the organizing principles behind these new urban designs and their publicity events. As old-style 'public space' declines and popular control of the streets becomes a thing of the past, a new-style 'publicity' or 'promotional space' evolves in which the reputation of the sponsoring corporation is visualized and its production of 'civic values' promoted. Let me turn to look specifically at Battery Park City and discuss the manner in which this new form of public/private space is exalted/exploited by the communications industry and how it positions us as spectators in scenographic events.

To begin to tell the story of how the public space of Battery Park became enclosed as a corporate domain, we start in 1966, when the Lower Manhattan Plan and its Rockefeller advocates noted that the financial district was running out of space in which to expand. They proposed that office and residential towers be built on landfill extending out to the pier line along the entire waterfront from the Brooklyn Bridge on the East River to Fulton Street on the Hudson River—six residential communities housing between 10,000 and 15,000 people were planned—each to be centered on a waterfront plaza located at the axis of Wall Street, Broad, Chambers and Fulton Streets. In order to carry out some of these plans, it was proposed to use the soil excavated from the World Trade Towers' construction site for landfill, creating what would become Battery Park City. In addition the state created in the late 1960s the quasi-public Battery Park City Authority, empowered to sell bonds in order to finance its landfill operations, remove the decadent piers and complete the necessary infrastructure. By 1976, however, Battery Park City was a barren space unable to attract development interest. The intervening years were, of

[14] Ibid, p. 89–110. The quotation appears on page 90.

course, exactly the years when New York slid towards bankruptcy during its fiscal crisis. Corporations and well-to-do residents began to flee from the city, its manufacturing base was eroding, its property values were faltering and its low-income neighborhoods were costing too much to sustain. In such financially gloomy times, Battery Park City Authority was unable to sell additional bonds and thus could neither raise the money for development loans nor attract development interest.

Since the fiscal crisis, however, economic development strategies have been focused entirely on taking advantage of the direction the global economy is moving in order to attract the headquarters of multinational corporations, global financial concerns and all the infrastructure and white-collar service support that these industries require. By 1977, New York City's office market began to recover, and its image as a glittering world city re-emerged. So the need arose, once again, for corporate support services, for communication services, for entertainment spaces and for luxury housing to complement this expanding white-collar employment. Now the World Trade Center was becoming a dominant force in the restructuring of Lower Manhattan. This new transportation, tourist and commerical hub was pushing the office center toward the west and the river—in other words, towards Battery Park City. To the northeast of the landfill, the old cheese, egg and milk wholesale district and manufacturing areas of Tribeca experienced a lively pace of luxury residential loft conversations. So it was argued that if the Master Plan for Battery Park City drawn up in the 1960s could be reorganized and allow more flexible and smaller-scale development and if the ideal of locating mixed income residential communities in Lower Manhattan could be abandoned, then the Authority believed the barren landfill of Battery Park City might appeal to local developers. The time seemed ripe for restructuring the view of waterfront development: disregarding the superblock megastructure mentality of the modernist 1966 plan, the revised Master plan of 1979 believed that conventional building lots and streets with sidewalks would enable the normal rules of Manhattan block and lot development to prevail. With these new plans in hand, developers responded and construction of a luxury enclave began.

The extent of this spatial restructuring by private developers utilizing public subsidies and support is massive and involves not only Battery Park City, but Riverwalk on the East River, South Street Seaport, South Ferry, Hudson River Center and Trump City plus a great esplanade from 42nd Street south to Battery Park City. And if we move inland it involves a new node of luxury at Union Square, a huge development at Times Square and in the outer boroughs projects just beginning on the Hunter's Point waterfront, the Brooklyn waterfront, the far Rockaways and inland a huge site on top of the Sunnyside railyards, not to mention Metro Tech, and so on bit by bit, node by node across the entire city. As a result of this spatial restructuring, New York displays the simultaneous characteristics of a 'dual city' with fantastic growth at the top of the economic ladder and abnormal decine at the bottom. Hence the ideological narratives which are imbedded within the current city image campaign selling 'New York Ascendant' become important legitimating procedures for the global economy that cover over an expanding gap which pits impoverished manufacturing workers and the squeezed middle class against increasingly well-to-do white-collar workers. If the dualities of wealth and poverty tarnish the city's liberal tradition, what better way to veil these discrepancies and release the tension they create than connecting to the present the civic tradition of New York's architectural heritage through fictional re-creations?

And so the creation of Battery Park City, a place that advertisement claims 'It's more New York, New York.' And Goldberger calls this triumph of urban design 'close to a miracle,' being 'the finest urban grouping since Rockefeller Center.'[15] This 'urban dream' is based on what he refers to as a radical notion—the 1979 master plan decided and most New Yorkers agreed that there was nothing wrong with Manhattan as it appeared and that 'most of the attempts by architects and planners to rethink the basic shape of the city have resulted in disaster.' But now the master plan is firmly in control of the look of the place, negotiating through its design codes the architectural styles, historical allusions and meaning of this period piece in return for giving a private developer permission to build.

Nevertheless this scenographic arrangement is a compositional form which explicitly relies on a series of familiar, non-disturbing and comfortable views taken from New York's architectural past: the look of the Brooklyn esplanade, Central Park West and Park Avenue are all re-constituted within Battery Park City. Here one can find the reproduction of Central Park lamp posts and benches, the inspiration drawn from the private enclave of Gramercy Park, as well as the great landscape inheritance of Olmsted's parks. Consequently, the images collected in Battery Park City become metaphorical carriers of a special kind of history and defenders of a set of values established in earlier times. Through the re-creations of traditional New York spaces and architectural forms, the present is filled with a sense of grandeur and self-importance or pleasure and excitement that many modernist places in the city apparently failed to achieve. Battery Park City teaches us how to feel, not think, about the past—how to overcome the sense of failure and crisis that modernism provided, that the near bankruptcy of the city's fiscal crisis congealed and the decline of American supremacy revealed. In a backward binding gesture, it stitches the production of this new luxury residential and commercial enclave to the architectural history of the city and thus establishes the illusion that this space of New York was always there, or at least its essential aura was never added nor artificially

[15] Paul Goldberger, 'Battery Park City is a Triumph of Urban Design,' *The New York Times*, August 31, 1986.

produced.[16]

The citation of these few images drawn from New York's commercial heritage seems to justify the entire outdoor museum of Battery Park City. As a predominantly commercial adventure, some might question whether the city and the state should be involved in helping to subsidize a fundamentally upper-class project. But then, this is the intention of nostalgia, to invert and gloss over reality. As the boomtown mythology of New York's heroic era of architectural and commerical development, a period that took place between the two world wars, is carried over into the 1980s and 1990s, Battery Park City's development energy might simply spill over and push the project northward from Chambers Street as far as Canal Street, where it could add another 68 acres. And some of the residential assignments might be transferred instead to more lucrative commerical development. And why should development stop there? For the entire waterfront around Manhattan, that fallow landscape of decaying piers and underutilized spaces which someone has described 'as being like an unhemmed dress,' this too might follow the path of tradition outlined in the 1920s and 1930s and be restructured and recycled with monumental commercial development.

These city images reappearing in the architectural creations of Battery Park City become situated quotations. They are public images appropriated for private means. If Battery Park City's $4 billion display of architectural might is contaminated with commercialism, it is after all the World Financial Center. And like all commercial architecture, its scenographic compositions invoke the art of advertising as well, either by providing the spectator with new landscapes of consumption such as the squares, cafes and shopping arcades specially designed to be visually entertaining or literally as consumers of images within these landscapes set up to attract attention. This has been called the pseudo-public sphere: where advertisers, public relations men/women, publicity agents—all manipulators of public opinion—intentionally insert their voices into the private realm of fantasy and imagination to spark the spectators's desire to consume or to instill an aura of goodwill in the art of forming consensus.[17] Not suprisingly, we find that the visual imagery and scenographic spaces of Battery Park City have been colonized by the creators of this pseudo-public sphere. The official opening of the Financial Center in October of 1988, for example, was accompanied by five days of promotional celebrations including boats, carnival floats, dance and musical performances along with the appropriate circus-like posters, buttons, banners and shopping bags.

A series of photographic essays entitled 'City Tales' was a fundamental component

[16] Slavoj Zizek, *The Sublime Object of Ideology*. New York: Verso, 1989, p. 104.

[17] Oscar Negt and Alexander Kluge, 'The Public Sphere and Experience: Selections,' *Oppositions*, 46, Fall, 1988, 60–82.

of the selling campaign (see Fig. 6.3). These 'City Tales' tried to create an image for Battery Park City meant to attract customers to the more than 20 shops, restaurants and cafes that lined the lobby corridors of the World Financial Center. In addition, by asssociating the look and the feel of well-established New York places and events with the newly created Battery Park City, these ads hoped to instill in the viewer the sense that this new landscape was none other than a natural node of New York, a dynamic place where people work, live, shop and eat, an exciting center teeming with art and culture and a public place that deserved recognition as well as a visit. These advertisements financed by the developers of The World Financial Center were placed in *The New York Times Sunday Magazine*, *Vanity Fair*, and *the New Yorker* and were to be twelve in number. After protests by the local retailers, however, that the advertisements failed to mention their shops, the campaign has taken on a more traditional bent. Nevertheless, the series of 'City Tales' that did appear reveal not only the type of 'public' that Battery Park City is designed for, but the legitimating rhetoric that its imagery instills. The 'City Tales' read something like this.

Poet Dana Gioia's 'City Tale of Destinations: Arrivals and Departures' selected Grand Central Station as her theme. While claiming that most contemporary travel feels more like commuting, she noted that one experience, that of Grand Central Station, still actually feels like travel as one enters into or departs from New York's greatest indoor public space. In this 'democratic precinct,' really a theater stage, one can meet people from all walks of life, each with their own compelling story. "This is a place which recognizes the importance of each arrival." Nevertheless not one recognition is made throughout this essay that Grand Central Station is simultaneously a magnificent landmark structure from New York and is a shelter for hundreds of homeless; that every arrival is greeted not only with the view of its cavernous lobby but with the outstretched hands of panhandlers and the wretched face of poverty; that a war over public space has its seamiest sites in the Station.

The differentiation of space

Let me leave Battery Park City behind and return to the general theme of architectural compositions for the rich and the poor in the contemporary city. A few years ago I began to describe the built environment of American cities as being controlled by a series of pattern languages: well-composed ornamental nodes generated from a set of design rules or pattern guidelines (like those put in place in Battery Park City). These ornamental fragments were planned or redeveloped as autonomous elements with little relationship to the metropolitan whole and with direct concern only for adjacent elements within each node. They are well-designed places of strong visual identity, special districts controlled by contextual zoning or design guidelines; shopping malls, festival marketplaces and theme parks whose visual decor and ambience are cleverly managed and maintained; and cluster developments of luxury housing, vacation retreats and retirement communities whose sense of place rests on well-articulated themes. These city places seem to have a serial appearance, mass produced in city after city, from already known patterns or molds. (Indeed some have claimed that a sure sign of Battery Park City's success is the number of look-alike developments that have sprung up in other American cities). But I want to specify this argument still further, for rather than seeing a uniform pattern of almost identical places being generated in cities around the world, and in spite of intentional rip-offs, now I find that designers of the built environment seem intent on detailing these places, specifying the unique visual quality and historic imagery that differentiates one place from another.

We might expect since capitalism shifts its form and mode of operations over time, then so would its spatial and aesthetic politics. In fact, many large metropolitan areas in the 1970s and 1980s have witnessed major spatial restructuring because differentiation between places has become increasingly important in the capital investment and relocation game.[18] Let us look at the reasons why differentiation of space has become the focus. During the 1970s and 1980s a new network of global cities arose, taking charge of coordinating the world-wide circulation of capital, goods, labor and corporations. The name of the game in the first-tier cities with global reach is to attract and retain the headquarters of multinational corporations and all the business services that these corporations demand, such as international banks, advertising agencies, legal, accounting and communication support.

But that is not all in this game of spatial roulette: computerized technology has enabled capital in the 1970s and 1980s to become increasingly flexible, able to move from place to place according to various locational preferences. And mergers have made it even easier for larger corporations under unified command to shift around various segments of their operations, facilitating cities of many different sizes both inside and outside of the United States to compete for flexible investments. As computerized information systems make white-collar service jobs increasingly portable, America's midsized, or 'second-tiered,' cities are experiencing fantastic growth—for these areas combine the advantages of both town and country, affordable housing and good jobs, yet still have room enough to accommodate that American urge to improve. A recent *Newsweek* report on these 'hot cities' noted that the site-selection committee of Sematech, a computer-chip consortium, reviewed the credentials of 134 cities before locating its

[18] David Harvey, *The Condition of Postmodernity*. New York: Basil Blackwell, 1989.

headquarters in Austin, Texas. Evidently in their quest for livability and good business climate, only two big cities, Boston and Kansas City, made it onto the list of Sematech's 25 finalists.[19]

City images become essential in this marketing game: the kind of image that spatial pattern languages can foster and sell. But marketing a city's image works both ways: industries can also enhance their products or services by association with a positive image of a city. Whether it is the city or the product that is for sale, surplus capital is drawn away from production in order to create consumer demand through the art of selling. Thus, in a not so surprising gesture that links the promotion of businesses with that of cities, forty large companies, including IBM, American Express and New York Telephone, recently formed 'The Alliance for New York City Business' and sponsored a pro-bono advertising campaign in print and on television that extolls New York City as a dynamic business center and an excellent place to live.

Now architectural associations are aware that cities must improve their image's if they are to play in the relocation game and that architects hold important restructuring roles in the development of style-of-life spaces and livable places. So, for example, the American Institute of Architects and the Royal Institute of British Architects conveyed a 'Remaking Cities Conference' in Pittsburgh during the spring of 1988. Their brochure advertising the conference seems to say it all: 'After more than 100 years, the industrial revolution is dying. Smokestacks are coming down; a global economy is emerging. Information networks, instant communications, and fast, efficient transportation allow things to be made where and when they are needed. People and businesses are no longer tied to places . . . Cities, the basic building blocks of industrial nations, are challenged by dramatic and rapid change . . . Businesses and individuals—increasingly free to locate where and when they want—select cities with the finest features and benefits. They look for history, culture, safe neighborhoods, good housing, shops and education, and progressive local government. Cities are competing, and their edge is livability . . . Livability is the new measure of cities. It is the qualitative scale on which they must compete for emerging opportunities . . .'[20]

In the competitive war now being waged among cities, style of life and 'livability,' visualized and represented in spaces of conspicuous consumption, become important assets that cities proudly display. But spatial restructuring engenders uneven development. As attention is focused on the upscale and livable urban environment, it is simultaneously withdrawn from impoverished and abandoned territories. Indeed pro-growth advertising

[19] *Newsweek*, Hot Cities, February, 1989, 42–50.

[20] Advertising Brochure for 'Remaking Cities' Conference, Pittsburgh, March 2–5, 1988. Convened by The American Institute of Architects and The Royal Institute of British Architects.

campaigns and locational incentives are invisible government subsidies: they are called private market endeavors, even though they target millions of public dollars to the upper classes and well-to-do sections of town and withdraw subsidies from the lower class and poorer neighborhoods. Property and income tax abatements for home ownership, for historic preservation and for the renovation of older structures which have underwritten the residential and commercial gentrification of historic and older areas of the inner city are not considered public subsidies even though they lower the city's overall revenue base.

Nor are corporate income tax abatements and infrastructure provisions considered to be the result of direct public policy, although they sweeten the prospect for private enterprise to invest project by project in the construction of new conventions centers, new sports arenas, new office complexes, new luxury hotels and residential structures, new retail and cultural centers. These are all written off in the rhetoric of economic vitality as market incentives that have helped private enterprise reinvest in the city.[21] But again this privatization of public issues bypasses the source of social conflict. It produces massive spatial restructuring which is the result of no overall public agenda or city plan and offers no political forum for debate. Information technologies may have enabled corporations to create a worldwide decision-making base and market for their goods, but they have simultaneously eroded the public sphere and discounted the process of public accountability.

In the interest of luring capital investment to our contemporary cities, architects and city planners are inserting many well-designed nodes or scenographic stage sets into the space of the city until a matrix or grid of these nodes develops which encourages horizontal, vertical and diagonal linkages that suppress the sequential order of reality, the connecting in-between spaces, and imposes instead a rational and imaginary order of things. These grids interrupt our vision of the whole; we focus only on the empty box which can be filled with any combinatory pattern and as a result we have no image of the contemporary metropolitan totality that might link us together, no map that spells out its pattern of uneven development and no plans for the city beyond the well-designed node. These are only some of the hidden costs embedded within 'The City of Illusion': the forgotten needs of interstitial spaces pushed through the sieve of spatial restructuring; the unmet education, health, housing and employment demands of the majority of people that make up the marginalized segments of the public; the inversion of terms such as 'public' and 'private' with the increasing privatization of what was once public discourse and public terrains. And these neglected spaces and forgotten needs, these inversions of discourse, of course, should shape our public architectural and city planning agendas as we move towards the 21st century!

[21] Michael Peter Smith, *City, State & Market*. New York: Basil Blackwell, 1988.

Six

MEGASTRUCTURES AND URBAN CHANGE: AESTHETICS, IDEOLOGY AND DESIGN

Darrel Crilley

Queen Mary College

Amid the dramatic reformation of the urban landscapes of New York and London, nothing epitomizes the emergent character of late capitalist space more graphically than the production of commercial megastructures. Dating from architect Raymond Hood's prefigurative 1929 vision for a city under a single roof (Koolhaas, 1978), through its embodiment in the development of Rockefeller Center (Tafuri, 1980), to contemporary renditions in the form of hermetic 'white islands' such as Detroit's Renaissance Center (Marder, 1985), the concept of 'total environments' that internalize the multiple functions of the surrounding city has a clear place in the trajectory of capitalist urbanization. What differentiates the contemporary period is the proliferation and aestheticization of such microworlds. As Christine Boyer argues in Chapter 5, this occurs within a fragmented metropolis, marked by the absence of any comprehensive or 'strategic' guiding vision for how the city as a whole may develop. Foucault, though speaking of the archipelago of disciplinary and institutional spaces, has used the term Heterotopia to capture this logic of a society where spatial dissociation and insulation become a common trend. The notion is easily extended to an urban context. As Davis (1988), among others, has noted, contemporary megastructures embody a dual logic: they are at once systematically segregated from the city outside—'gigantic anti-urban machines' in Tafuri's (1980) phrase—and also 'claustrophobic' space colonies attempting to recreate 'the genuine popular texture of city life' within themselves (Davis, 1988, p. 86).

The heterotopic tendency reaches its apotheosis in the two microworlds considered in this chapter, New York's World Financial Center (hereafter WFC) and London's Canary Wharf. Grandiose set pieces of the type advocated by theorists of a 'Collage City' (Rowe and Koetter, 1978) they are designed as autonomous entities, with concern only for what is immediately adjacent or directly contributive to their rental value. As one planner of Canary Wharf explains, the task was 'grafting into London a whole new piece of city, in an area where we don't have to destroy buildings to do it' (Olympia and York, 1990a, p. 8). Standing authoritatively on the skylines of both cities (see Figs. 6.1 and 6.2) and comprising, essentially, 8 to 12 million square feet of office space, these are among the most conspicuous testaments to finance capital's ability to dictate the rhythm and form of urban change. They are shaped from the same processes: both are driven by a speculative imperative to accommodate financial services restructuring; are managed by a globalized developer (Olympia and York—hereafter O&Y); are designed by an international entourage of architects; commissioned by local and central governments prepared to offer public subsidy and support; and, especially in the case of Canary Wharf, attempt to create a 'Complete Urban Community' fusing the 'best' dimensions of 'city' and 'country-town' life into a stimulating new urban lifestyle (O&Y, 1990a). These centers for high finance are differentiated only by timing, institutional form and variations in outward appearance. Intended as key nodes in globalized financial markets, they are fragments of an increasingly standardized postmodern landscape that, rhetoric of regionalism apart, fast begins to exhibit the same serial monotony of which international style modernism has been accused of.

This basic similarity stems from their production by a developer with global reach. O&Y acts as an international mercenary, intensifying inter-urban competition by rapidly dispatching the infrastructures demanded by all cities that vie for dominance atop the urban hierarchy.[1] Armed with massive private financial resources, the company exhibits no particular loyalties to the eventual supremacy of one city over another. To borrow a metaphor from Smith (1984), O&Y is like a plague of locusts, settling on one city only to move on and devour the rich package of incentives that competing city growth coalitions feed it. Thus in 1987, while O&Y's architects were adding the finishing touches to the WFC and installing a multimillion dollar arts and events program, their chief executives were making deals to take on the Canary Wharf venture in London. And both of these projects were preceded by the lead O&Y took in the rash of speculative redevelopment in downtown Los Angeles during the early 1980s.

The unity of the schemes is further sealed by the iconic status of both as glittering jewels heralding renewed downtown vitality and a positive reversal of urban fortunes.

[1] Olympia and York is one of a growing number of globalized property developers with a portfolio spread across North America and Europe, with outliers in Tokyo and Moscow. Privately owned, its property assets in 1990 were conservatively estimated to be worth between $11bn and $18bn. It uses its basis in property investment to diversify into other sectors: it has controlling interests in the production and distribution of Canadian energy; it controls Albitibi Price, the world's largest newsprint producer; and until recently had significant stakes in North American retail and transportation conglomerates. Canary Wharf, its largest development to date, is estimated to involve investment totalling $6.6bn, of which the first $2bn emanates from O&Y's own funds. It is this ability to avoid institutional sources of finance that enhances their resiliency vis-à-vis competing developers. For assessments of O&Y see for example: 'Inside the Recihmann Empire,' *Business Week*, January 29, 1990, pp. 32–36; 'Reichmann Brothers: Kings of Office Land,' *Economist*, July 22, 1989, pp. 17–20; 'An Empire Built on Property,' *Accountancy*, vol. 104, no. 1152, August 1989, pp. 107–109.

Granted a privileged status above mere office blocks, they are represented—by their enthusiasts, architectural commentators, real estate pundits and national media—as multiple successes. Politically, they are touted as symbols of the continuing ascendancy of New York and London to the status of command and control centers within a reconstituted

Fig 6.1. Pelli's skyscraper at Canary Wharf, out of scale with London's cityscape.

world financial system. One architect at Canary Wharf, for instance, views his work as a 'development for the third millennium,' the shimmering steel surfaces of its skyscraper portending an optimistic and expansionist future for London (Pelli, 1989). Behind this success, conservative commentators see 'a thriving symbiotic union between the public and the private sectors' (Gill, 1990, p. 74). Further enthusiasm derives from the role each project plays in underpinning two large-scale redevelopment initiatives that have proceeded in tandem throughout the 1980s, namely, the 'renaissance' of London's Docklands and 'triumph' of Battery Park City (Deutsche, 1988; Zukin, 1989). Whereas the latter is made viable by the economic and symbolic gains accruing from construction of the WFC, Canary Wharf invests the former with a mythology of national revival and prosperity (Crilley, 1990a). More locally, 'public' gains in the form of 'linkage agreements' and new forms of public space are listed among the triumphs of the new megastructure. Goldberger (1988), for instance, deems the WFC's glass-enclosed 'Winter Garden' and Plaza to be the most majestic public spaces built in New York since Grand Central Terminal, reversing the impoverishment of the public realm in the face of privatization. Likewise, Sir Roy Strong (1988) defended Canary Wharf as 'more truly in the idiom of this metropolis than any for a very long time.' Aesthetic achievements also rank among the benefits of megastructures. In particular, their design is celebrated as the final death knell of an authoritarian, ascetic modernism and the flowering of a humanist postmodern aesthetic. Thus, as urban design, they are deemed the high point of contextualism, replicating the 'success' of Rockefeller Center in 'weaving' themselves into the 'fabric' of the city (Dean and Freeman, 1986). And, as architecture, both projects are hailed as positive contributions to the development of skyscraper form (Goldberger, 1986; Pelli, 1989).

Fig. 6.2. The World Financial Center (left foreground) on the New York skyline.

The intention in what follows is to counter these dominant representations through a critical analysis of the roles that aesthetic practice and discourse play in the development of the megastructure. Following Deutsche (1988), it is possible to read WFC and Canary Wharf not as signs of enduring vitality, but as enormous and cautionary symbols of changes underway in the relationship between property development and aesthetics. In keeping with Deutsche's strategy, it is possible to brush the documents of urban ascendancy 'against the grain,' demystifying their ideologies and reinstating their 'repressed texts.' Thus, to counter representations of Canary Wharf and WFC as benevolent private initiatives fulfilling essential needs of world cities, I begin by challenging this ideology of utility, pointing to their competitive, speculative nature and deep dependence on public support systems, both legal and financial. Against an aestheticizing view of both projects as triumphs of beautiful, livable environments I show how the coterie of design professionals and cultural actors involved eschew political and ethical considerations to adopt a position of 'corporate moral detachment' (Wodiczko, 1986) and wholesale cooperation with the forces of power. Relatedly, it is stressed that the humanistic claims of architects and designers to meet public needs can only be sustained by denials of the low-income housing and employment needs that go unmet because of the subordination of the city to the logic of exchange value. Far from being neutral guarantors of democratic legitimacy, these actors are active ideologists, mobilizing 'meaning' in both built and textual forms to provide redevelopment with an acceptable cultural alibi. Thus, to foreground what follows, while designers emphasize the rescue of public space in these projects, this emphasis masks the exclusions inherent in their construction of a homogeneous 'public' and the removal of land-use and design issues from public control. Likewise, while both projects entail dramatic social and physical upheavals as part of the 'creative destruction' of late capitalism, this is papered over by historicist designs that stress principles of cultural continuity, stability and permanence. And, concomitantly, just as both projects realize Lefebvre's (1979) misgivings about 'abstract space'—thoroughly commodified, uniform, universally divided into interchangeable parts, hierarchically ordered into center and periphery—each development is represented as the epitome of urban place making by a familiar postmodernist discourse stressing variety, visual detail and local idiosyncracies. The broad motivation for what follows, therefore, is to mount an analysis and critique of the modalities by which the architecture and design of these two projects function ideologically. As such, it is a demonstration of the ultra-conservative political associations of certain strains of postmodern design in relation to the production of the city. Grasping the instrumentality of aesthetics requires, however, prior specification of the political-economic context in which they are situated.

Urban political economy: The myth of the local hero

> *It started with urgent needs—London's needs. The need for a new supply of fine modern premises, with large floor areas and high ceilings; the need for new single buildings to house the headquarters of important international companies; the need for greater economy in accommodation costs; the need for higher quality in materials, in systems, in services, in amenities and in the surrounding environment.*
>
> O&Y, 1990a: advertisement for Canary Wharf

Thus do O&Y portray themselves as local heroes who think long-term about servicing the city's needs and who rescue redevelopment agencies from bankruptcy and failure. Both the commercial zone of Battery Park City (for which they assumed complete responsibility to build all commercial phases, in a guaranteed record time of five years) and Canary Wharf (which they took over when the original developer (Ware-Travelstead) withdrew support from the project in 1987, saving the Thatcher government from political embarrassment) engender a view of them as saviors of the city, shouldering high-risk projects in 'derelict' locations that less 'adventurous' and 'daring' developers abandon. Optimistic tropes of the type 'what is good for O&Y is good for the city' abound. For New York State Governor Cuomo, WFC represents 'a soaring triumph' for Wall Street, ensuring New York's competitiveness as a headquarters location for financial services (Gottlieb, 1985). Guardians of London's interest find similar salvation in Canary Wharf. For one planner it simply represents 'London's new heart transplant' (Robson, 1990, p. 10). According to Lord Young, former Trade Secretary, it betokens a 'future that works.' As he reflects on his own ratification of the scheme: 'London is bursting at the seams and we have to find somewhere to accommodate that growth. The only way to go is east. That is what the whole of Docklands, not just Canary Wharf is about . . . the whole balance of London will change . . . [and] the effect will be to maintain London as the second financial centre of the world and to consolidate it as the financial centre of Europe' (quoted in O&Y, 1990b). More optimistic still are recurrent parallels between the strategic significance of London's former docks in the development of Britain's empire and trading hegemony and contemporary undertakings at Canary Wharf (cf. Crilley, 1990a). Thus, a recent plea for yet more subsidy for Canary Wharf is aptly titled 'Building up the nation's prestige' (*Estates Times*, 7/20/90). Canary Wharf, it enthuses, is a much-needed 'replacement' and 'extension' to the decaying and inadequate 'trading' facilities of the City of London that will begin to 'redress the imbalance' between the three main trading centers of London, Tokyo and New York.

Along with this essentialist view of Canary Wharf as inherently functional and as a simple reflex to disequilibria, there is optimistic faith in a 'growth is good' ideology and the inevitability of 'trickle down.' Like the wider growth ideology of which it is a part—variously termed renaissance, regeneration or revitalization—this consensual image of Canary Wharf as an embracing benefit for the collectivity of London disavows the drastic unevenness of redevelopment processes of which it is a prime constituent. Developments of this size, so the argument runs, cannot fail to generate beneficial effects for surrounding low-income populations even if they do not gain directly (Church, 1988). Moreover, as O&Y proudly declares, it is working for and liaising with 'the community,' positing itself as a 'corporate citizen' (O&Y, 1990b). It has signed accords with local government in London and participates in a linkage agreement in New York, generating funds for housing rehabilitation in the latter under the aegis of the Housing New York

Corporation and guaranteeing 2000 construction jobs for 'local' people at Canary Wharf as well as sponsoring training initiatives to rectify the skills mismatch.[2]

Such a legitimating repertoire is itself a reflection of the volatility and instability that grips city governments in an area of heightened capital mobility, with its attendant threat of fiscal crisis (Harvey, 1989c). This defensive, hostage mentality leads to government policies, originating both centrally and locally, that patently serve the particular needs of profit at the expense of general, social needs. In this sense, exhortations to utility and the general good become ideological, dissembling the unequal nature of the public-private partnerships involved and deliberately distorting the reigning priorities (Squires, 1989).

Far from being a bold benefactor, O&Y operates on terrain prepared by an extensive network of public supports, ranging from direct financial inducements, through infrastructural provision to measures insulating it from public influence. There is striking convergence between the modalities of public subsidy adopted in New York and London. In New York, O&Y benefitted from Mayor Koch's manipulation of the tax code to stimulate private development. Despite oppositional claims that O&Y, in view of their Manhattan property empire, did not need or deserve tax abatements, Koch forced through an abatement package such that the City would realize no taxes from the WFC for ten years, thus foregoing $117 million (*New York Times*, 10/16/1981). At the same time, O&Y received off-site benefits through multiple tax reductions on six of its New York properties totalling $98 million. These circumvented the usual procedures for tax assessments, being sanctioned by the finance rather than the tax commission. Irrespective of the confusion of legal responsibilities however, the tax commission had already awarded O&Y additional reductions of $45.5 million (*New York Times*, 8/24/1988). Meshing with this, by investing at Battery Park City, O&Y was able to escape the usual democratic procedures governing planning and land-use decisions in New York. Because legal ownership had been transferred to the New York State Urban Development Corporation (an unelected, unaccountable body—'corporations without shareholders, political jurisdictions without voters or tax payers' (Walsh, 1978, p. 4)) the city

[2] All such planning gain deals are beset by a number of difficulties that undermine their credibility as solutions to urban problems: the secrecy of their signature; the disparity between meager community gains and vast developer profits; the uncertainty of gains flowing from volatile real estate markets; difficulties of legally enforcing obligations to provide employment for locals; and again, the short-term approach they encourage to metropolitan strategic planning. For example, while O&Y have made substantial progress with training initiatives to employ local residents in the construction phase, this in no way obligates future tenants to employ locals. Moreover, this was traded off against the right to public inquiry and Tower Hamlets Council's agreement not to contest Compulsory Purchase Orders used for the construction of the Docklands Highway servicing Canary Wharf. Moreover, the promise of a £2.5m training package represents just 0.06% of Canary Wharf's estimated £4.4bn value at 1989 prices and can scarcely be considered a fair financial gain relative to the vast public funds underpinning the project. Aside from these specifics though, the most serious problem with 'linkage' is that it encourages a view of redevelopment as a solution to housing and employment problems when it is, in fact, both directly and indirectly (through its inflationary impact on land and housing markets), a cause of housing needs crises in both New York and London.

relinquished its legal control over the project's course and along with that nullified existing public review processes—community board reviews, public hearings and City Planning Commission consent.

At Canary Wharf, public support is even more lavish and transparent. Sited mainly within an Enterprise Zone administered by a non-elected, non-accountable central state Urban Development Corporation (the London Docklands Development Corporation), Canary Wharf was absolved from any substantial need to comply with local borough strategic plans and removed from any form of public inquiry.[3] As one local politician quipped, one of the largest real estate ventures in the world was subject to less scrutiny than a 'planning application for an illuminated sign on a fish and chip shop in the East India Dock Road' (DCC, 1988, p. 8). Under the guise of antibureaucratic efficiency, this abrogation of democracy permits a corporatist style of piecemeal deal-making to go on behind closed doors. The development of Canary Wharf is governed by a 'Master Building Agreement' signed between O&Y and the LDDC in 1987 in a vacuum of strategic thinking, without conventional assessment of its environmental and social impacts and in a secretive atmosphere where the public has no rights extending beyond the benevolence of the developer and a non-plenipotentiary position in LDDC planning committees. Thus, despite the profound strategic consequences of Canary Wharf, the process governing its approval involved chronic myopia viewing it as just another 'project.'

Enterprise Zone status also confers on O&Y significant financial gains. The frenetic building boom that has transformed the Isle of Dogs in the past decade is not driven by any wish to fulfill unmet demands, but by a system of tax incentives and capital allowances that merely heightens the tendency to speculate in property as a pure financial asset (DCC, 1988, 1990; Haila, 1988). Not only did O&Y receive a gratuity in the form of 20 of the 72 acres of Canary Wharf at a discount price of £400,000 per acre when estimated market prices on the Isle of Dogs stood at £3 million per acre (National Audit Office, 1988), but more significantly, assuming all 12 million sq. ft. are developed, there will be a disguised public expenditure in the form of £1.33 billion deferred capital allowances (DCC, 1990). It is this fiscal incentive, scarcely mentioned in the euphoria surrounding the project, that underpins it.

Likewise, though euphemized as 'leverage' or 'just-in-time infrastructure,' the current suite of transport initiatives in Docklands (construction of the Docklands Highway, extensions of the Docklands Light Railway into the City of London and a new subway line), though financed by public-private consortia, is targeted exclusively at making

[3] Broader accounts of redevelopment in London's Docklands and why it has proved so contentious can be found in Brownhill (1990), Smith, (1989), Brindley et al. (1988) and the Docklands Consultative Committee (1990).

Canary Wharf a viable business location at the expense of strategic transport planning for the metropolis as a whole (Church, 1990; DCC, 1990; Brownhill, 1990). These initiatives are, in fact, responses to persuasive lobbying by O&Y, who utilized the Master Building Agreement to extract infrastructural gains tailored to their priorities. In the case of the Jubilee Line, for instance (itself a version of an express rail proposal emanating from O&Y in 1988), alternative transport schemes are suspended while the government foots the bill for a project whose rapidly escalating cost may consume £3 billion of public funds. Meanwhile, the contribution of O&Y is fixed at £400 million. Both the routing of the line and direct connection between central government and O&Y mark this as a transparent show of support for grand property speculation, hedged by the political exigencies of securing Canary Wharf as a successful symbol of enterprise culture. As one recent commentator suggested: 'so naked is this gesture of support for the billionaire Reichmanns and their Thatcherite flagship on the Isle of Dogs, that even a fragile figleaf of concern for inner-city regeneration, in the shape of stations en route in rundown Southwark and Bermondsey, has been suspended' (*Building Design*, 12/10/90). It is a cynical deceit to view this as limited public funds leveraging private investment; rather, private speculation has cajoled massive public expenditure to provide a single developer with what it boasts of as its own 'purpose-built transport system.' In this process of 'reverse leverage,' an escalating quantity of public finance has been diverted into correcting the inefficiencies created by an ad hoc approach to creating the infrastructural web that would integrate Docklands with the rest of London. Moreover, the financial burden of forcing Canary Wharf through is placing fiscal strain on the LDDC. Its finance from land sales having been reduced to a trickle, the LDDC registered a deficit for the first time in 1989/90. But, owing to the political urgency of improving access to Canary Wharf, it had committed the bulk of its budget to a £640 million road-building program and a £240 million extension of the Docklands Light Railway. The underside of this frantic attempt to rectify inadequate infrastructural planning may once again be marginalization of social programs and a deepening of the unevenness of redevelopment even within the officially designated Docklands area.[4]

The public risk and political patronage that featherbeds O&Y is thus disguised by the neo-classical rhetoric of a self-regulating property market. This also contradicts the monopolistic drive of O&Y and masks the conscious intra-urban competition involved. Faced with a ten-year amortization time for a £4 billion investment, O&Y must confront its extreme 'local dependence' (Cox and Mair, 1988) and forge, in a highly aggressive,

[4] In 1989, for instance, the LDDC received a windfall gain of £359m, effectively trebling its budget for 1989–90. In 1990–91 it accounted for 61% of the total Urban Development Corporation allocation and 40% of the entire urban block. Despite this, programs for social housing have recently been suspended owing to prior designations for infrastructural provision.

interventionist manner, a mix of socioeconomic and physical infrastructures complementary to its key investment. These ameliorative measures assume a variety of forms. In both London and New York, however, the thrust is towards monopolistic control of local office markets. Replicating techniques it used in Manhattan in the 1980s to draw tenants into the WFC (purchases of existing buildings and leases, rents at half the cost of core locations, customized finishes), O&Y has attracted tenants from City landlords and even negated existing tenancies in Docklands. Thus, Merrill Lynch, the securities dealer, was enticed to relocate to Canary Wharf not long after leasing new premises in the City, O&Y taking responsibility for the 20-plus years remaining on Merrill's lease. Similarly, the Daily Telegraph building on the Isle of Dogs has been purchased and the company rehoused in Canary Wharf's skyscraper. Further leverage is gained by equity investments in competing property companies, thus giving it a sleeping role in developments in Docklands and other parts of London. The first and most significant of these was the purchase of one-third of Stanhope Properties in May 1988 and this was followed by gain of an 8.25 percent stake in Rosehaugh. At the time, these two companies were linked through Rosehaugh Stanhope Developments, being responsible for the Broadgate office complex (the largest City of London office complex built in the 1980s), the partially constructed Ludgate Hill project also in the City of London and, scheduled for the mid-1990s, the redevelopment of Kings Cross, a commercial undertaking of comparable scale to Canary Wharf. Through such tactics O&Y is maneuvering into a position where its standing in London will mirror its power in Manhattan as the city's largest commercial landlord. Potentially, it can influence the three largest London property projects for the 1990s, shuffling companies from one location to another.

Moreover, in pursuit of the mix of activities requisite to the concept of a whole new business district, it has spread its tentacles in Docklands. It has bought sites and entered into joint ventures on land adjacent to or near the project. Immediately north of Canary Wharf at Port East it is developing a retail and leisure complex, hoping to attract a major cultural institution, while immediately south, it has tied up with Docklands residential property specialist Regalian to provide a residential development. East of Canary Wharf, in an attempt to establish the international pedigree of its site, it has sponsored the construction of a 'China City.' Rationalized by the concept of 'critical mass,' these subsidiary investments are all geared towards diminishing the speculative nature of the project. However, it is doubtful if even a self-interested O&Y can compensate for the chaotic overinvestment and overproduction of office space that has occurred in the absence of any strategic coordination. Not only is the Docklands office market itself troubled with high vacancies, bankruptcies, devaluation of assets and frantic attempts to switch property between use classes, but the City of London has a record quantity of office space and a record vacancy rate of around 20 percent (Savills, 1990). This overproduction has to be seen as the product of the fragmentation of the metropolis into a patchwork quilt of competing intra-urban growth coalitions, each eager to establish or consolidate its status as a financial district. When Docklands gained Canary Wharf, for example, there was a consciously chosen strategy in the City of London to weaken conservation regulations, raise plot ratios, increase the level of permissions and generally pursue anything that would ward off competition from Docklands (King, 1990; Duffy and Henney, 1989). Similarly, though O&Y may eventually control Broadgate and Kings Cross, both these projects are not complementary expansions but competitive ventures. Whichever particular interest suffers loss, the 'lawless caprice' of deregulating property development in London stands to be regulated after the fact by devaluations and a game of musical buildings reminiscent of the boom-bust scenarios afflicting 'free enterprise' cities such as Houston and long characterizing New York's property markets (Feagin, 1988; Zukin, 1989). In fact, though WFC may now be successfully let, such that O&Y can contemplate sale of its assets to generate capital for Canary Wharf (Hylton, 1990), it is an integral part of the escalating vacancy rates besetting Downtown and Midtown commercial markets (*Financial Times* 6/15/90).

The organization of architecture and design

The task of fashioning the mental and physical representations of O&Y's megastructures resides with an interdisciplinary team of professionals comprising urban designers, architects, public artists and landscape artists. As practitioners within the built environment, they participate in the purposive production of meanings, forms and symbolism for that environment, mediating, together with advertising and architectural commentary, perceptions of socioeconomic transformation in the city (King, 1988). Like Rockefeller Center before them, Canary Wharf and WFC condense multiple changes underway in the patronage, philosophy and practice of architecture (cf. Tafuri, 1980). Discussion of these changes provides a glimpse as to how and where cultural producers contribute to the production of the built environment. It is a necessary prelude to understanding how design professions engineer consent for redevelopment.

Both these projects are illustrative of the heightened value attached to 'designed,' individualized products in a competitive market culture (Zukin, 1991). They epitomize a search for distinctive, novel buildings that will lease faster, for higher rents and for a longer duration, yielding the eventual occupier a headquarters rich in the 'symbolic capital' accruing from possession of a building designed by a 'name' architect. Moreover, as in Bourdieu's (1984) original conception of symbolic capital, signs of luxury and cultural discernment can be liquidated and converted into economic capital. Signature buildings such as the AT&T headquarters retain their value when sold to subsequent tenants and 'good design' (the elusive quality and buzzword among real estate developers), though initially more expensive, can ensure the product a longer economic lifetime. Thus, O&Y's brief to Cesar Pelli to make WFC a 'timeless' building with all the permanence connoted by a by a granite exterior and opulent marble interiors, was not an affirmation of the enduring validity of modernist aesthetics, but an attempt to emulate the continuing economic success of other iconic modernist buildings in New York, most notably the Rockefeller Center and the Seagram building. Though both these buildings were comparatively expensive to construct, they have continued to have high occupancy rates and command high rentals through nearly half a century of volatile real estate markets.[5] It is this search for longevity allied to fashionability that justifies the heavy investment in aesthetics at the WFC.

Design, in all its phases—from yielding visualizations and simulations of urban form to decorating the final structure—has become much more thoroughly integrated into the marketing of real estate. However, while some developers display particular stylistic preferences, O&Y shows a pronounced aesthetic indifference, sponsoring Cesar Pelli's neo-modernist design for WFC just as easily as it backs postmodern revival styles at Canary Wharf. Its patronage is not governed by a postmodernist desire for a more meaningful, communicative architecture, but rather by a dual concern to be up-to-date and employ those design 'superstars' authenticated by architectural criticism. The cultural validity of the building, and hence its marketability, clearly resides as much in the credentials of the architect as the stylization of the product. In this sense, the choice of Cesar Pelli for the WFC is an ideal cultural investment; his design is endowed with the consecration of 'high' architectural culture and popular recognition. As well as being acknowledged by the gatekeepers of architectural culture as one of the most legitimate practitioners of the skyscraper art, he carries academic credentials as former Dean of Architecture at Yale and celebrity status through hagiographic coverage in popular journalism.[6] With a reputation based on design of cultural institutions such as the New Museum of Modern Art in New York, Pelli is predisposed to imbue WFC with a cultural pedigree quite apart from specific judgements made of the building. By sponsoring an architect recognized as legitimate, O&Y clearly hope to gain broad public acceptance.

Moreover, Pelli's constellation of four buildings, by making an indelible imprint on the skyline, is likely to enhance the rentability of the office space within. At Canary Wharf, however, O&Y learned that employing a single-design architect has its practical

[5] For an architect's assessment of how to provide fashionable buildings with enduring rental values that replicate the Seagram building, see Burgee (1987).

[6] Celebrations of Cesar Pelli in the business press can be found, for instance, in 'Big Yet Still Beautiful,' *Time*, September 24, 1990, pp. 98–100, and 'The New Master Builder,' *Newsweek*, August 4, 1986.

and symbolic limits when it comes to attracting several tenants from the world of high finance: difference and variety is necessary to cater to the 'taste cultures' of business. Accordingly, in a maneuver that compresses the current explosion of styles within architecture into a single project, Canary Wharf involves a representative inventory of big-name architects from I.M. Pei through to Aldo Rossi. Most of the practices commissioned originate in America and were selected for their known ability to produce distinctive images and work within the rigid time constraints of rapid commercial development. Thus, Cesar Pelli, Kohn Pedersen Fox (KPF), I.M. Pei and Skidmore Owings and Merill (SOM) have all been selected to do at least one of the scheduled 24 buildings in preference to British architects, two of the practices using their involvement as a launching pad to establish European offices in London.[7] In response to self-defensive criticism from British architectural circles, however, high-style British practices have been given opportunities. Thus, the up-and-coming practice Troughton McAslan has found a place on account of its modish revival of expressionist modernism and Norman Foster has been called into design a complementary tower to Cesar Pelli's central 850-ft. skyscraper. Even more academic architects have been enlisted. The austere, rationalist design of Aldo Rossi, for example, was sought following his award of the Pritzker prize—architecture's supreme title of international cultural nobility. Each architect here vends his or her own style and 'ism' in an attempt to simulate the diversity of a historically developed city within a compressed time frame. The diversity is, however, scenographic, each practice working within the guidelines set by 'master' planners SOM to hang their design motifs on a basic structural frame.

Thus, the steel clad obelisk chosen by Cesar Pelli fronts onto the contextualist elevations of KPF's granite clad building; SOM's revival of neo-Edwardian classicism confronts Troughton McAslan's rendition of the wraparound granite and glass skin found on Frank Lloyd Wright's Johnson Wax building; and Aldo Rossi's abstract, geometric forms will adjoin the sweeping crescents of West Ferry Circus, designed by SOM to resonate with the grandness of London's Regency crescents. Though avid promoters of architectural pluralism such as Jencks (1987; 1988) might interpret this heterogeneity as a reconciliation between warring architectural factions of late-, post- and neo-modernism, its more telling significance should be that all architecture can be reduced to easily marketable styles and images by the levelling forces of the market. Nor should the notion of a pluralist architectural collaborative be conflated with the idea of political pluralism: to advocate stylistic tolerance in art and architecture does not mean they produce 'democratic signifiers' (Foster, 1987, p. 2). On the contrary, when the overarching

[7] Both Kohn Pedersen Fox and Skidmore Owings and Merill now operate London offices, a reflection of the fact that large U.S. practices now account for the bulk of large commercial work underway in London.

institution is the market, pluralism occludes a basic conformity wherein 'the difference represented . . . is the difference between a Ford and a Chrysler. Every commodity has to have its look' (Crimp, 1989).

More cynically, given that the architects involved now actively market and 'position' their designs, they are themselves responsible for reducing architecture to a form of consumer packaging. Monographs, newsletters and exhibitions all advertise the styles of the architects involved, aiding O&Y's in-house committee of architectural advisors in their selection. As architect and critic Kieran suggests, in an age when architectural practice is rapidly being reduced to façadism, propagation of a distinctive style is of the essence:

> *Given the competition between a relatively large number of architectural firms for a limited number of projects, the market economy has created unprecedented demand for image differentiation within the profession itself. At least at the highest levels of practice, uniqueness has become a prerequisite for survival. Since the core service performed by all architects is essentially the same, differentiation must be achieved in the secondary, formal realm. Packaging, style, special optional features, brand names and overall quality can be manipulated to establish specific, identifiable position within the marketplace.*
>
> (1987, p.107)

Furthermore, superficial stylistic equations of particular buildings with a single architect misrepresent the nature of the architect's involvement. In line with predominant trends in the practice of architecture, architects at Canary Wharf have been denied a position of comprehensive responsibility for coordinating the myriad pragmatic and aesthetic dimensions of building. Rather, they are exterior finishers 'decorating the shed,' to use Venturi's terminology, working within a complex professional and technical division of labor. In both projects, the role of architects is tightly and precisely circumscribed, relegated to finding suitable images with which to envelop the basic structure. By the time Cesar Pelli and Associates were brought in, their design task was confined to decorating the lobbies, the treatment of façades and culminating gestures on the skyline. This also applies to the series of lower buildings at Canary Wharf. Architecture is confined to surface treatments and extending the technology of cladding systems, whether to cultivate the sleek, unadorned façades characteristic of Troughton McAslan (for instance) or to achieve the ornate symbolism of SOM's Edwardian Revival building. To be sure, while critics such as Jameson (1984) may want to read a 'new depthlessness' into this culture of the image, architecture has been troubled by the lack of necessary relation between interior and exterior since at least the last century. What is distinctive is that, under the guise of rethinking modernism, architects have conspicuously ceased to talk about structure at all, openly embracing the skyscraper as an enormous package to be giftwrapped according to the client's taste. As Huxtable (1984) mockingly suggests, while architects may be generating a discourse on the art and virtues of large commercial buildings, 'having a wonderful time rearranging the deck chairs on the Titanic' (p. 101), the reality of their practical position has been to indulge in 'the vacillation of the façades' (p. 101). Yet, under a stylistic rhetoric of postmodernism and a hegemonic semiological approach to architecture, façadism has become one popular path to the salvation of architecture as a professional discipline (Crilley, 1990b).

The architects at Canary Wharf are presentative in this regard. They participate in a network of 'associations' and subcontracting, in which 'concept' practices concentrate on the design tasks while 'production' architects churn out the working drawings and detailed specifications that make the building technically feasible. Just as massive, integrated firms grew in the 1950s and 1960s to deal with the increasing size and complexity of building tasks, so now does disintegration and specialization, by function and style, yield an alternative way for architects to stake a claim in the production of the built environment's largest structures (Gutman, 1988). This mode of architectural production bears strong similarity to the nineteenth-century Beaux Arts approach in which pure design from the studio is handed off to production workers, whose drawings are in turn passed on to contract administrators (Brain, 1989). This form of organization—trendy design stars propped up more or less independently of production—is increasingly becoming a preferred mode of practice and lies behind the prolific output of numerous postmodern gurus such as Venturi, Rossi and Johnson. Certainly, this is not a universal mode of practice at Canary Wharf. Large integrated practices such as SOM and KPF can still deliver comprehensive packages to time and cost; but they also keep in-house stars who evolve individual identities as designers (Bruce Graham and Adrian Smith at SOM stand out as notable examples). Nevertheless, O&Y has a preference for adopting the Beaux Arts approach, purchasing artistic imagination from its name architect and technical efficiency from production specialists. Thus, at Canary Wharf neither Pelli nor Troughton McAslan manages the detailed phases after design; they are passed onto Adamson Associates, a Toronto-based practice specializing in the production of working drawings.

Cesar Pelli Associates is, in fact, a front-runner in founding its reputation on pure design. It has made a science of interviewing and cataloguing other architectural practices with which to combine. By associating with multitudes of other practices, each of which has its own specialties, his practice is able to retain the small size of an atelier but take on a growing volume of lucrative skyscraper commissions. It is inconceivable that Pelli's firm would have become so economically successful without adopting this stance. It allows it to put together architectural teams capable of winning commissions for large-scale projects from larger practices such as SOM.

While architects were previously criticized for being too demonstrative, authoritative and naively utopian in their aspiration to effect social transformation through architecture (Wolfe, 1981; Blake, 1976; Holston, 1989; Saint, 1983), the contemporary moment is disquieting for the opposite reason: resigned abandonment of social concern in the bid for professional status and individual recognition. The portrayal of Canary Wharf and WFC as selfless acts of cooperation amongst conflicting architectural views and as interdisciplinary collaborations with public artists only masks this lack of ethical focus. As Deutsche (1988) argues, what is forsaken in projects such as Canary Wharf is not egoism and self-expression in deference to team work and the use of a 'common language,' but ethical considerations. Postmodernist discourse, with its barely concealed agenda to resurrect architecture as a three-dimensional art form, the mode of practice with which it is associated and main currents of criticism facilitate, this shift from ethics to aesthetics (Huxtable, 1984; Ghirardo, 1987; 1990; Harris et al. 1982). Such trends contrive to ghettoize architecture, narrowing the concerns 'proper' to architectural discourse to the formalistic and symbolic. In such a context it becomes illegitimate to ask troubling questions about the social and political purposes of a venture such as Canary Wharf, let alone its disruptive social and strategic impact upon east London. Taking professional refuge in the realm of pure design, architects at Canary Wharf epitomize this tendency to eschew questions of politics in order to build glamorous and lucrative businesses. Who builds what, where and for whose benefit are all clearly external to the 'meaning' of architecture at Canary Wharf. Thus, despite Aldo Rossi's program to restore significance and meaning to architecture through the rational development of a language of forms and typologies that is insulated from the instrumentalities of political-economic power (Rossi, 1982), his architecture at Canary Wharf is reduced to a freely available object of consumption. It signifies not the autonomy and freedom of the architectural creative act, but its unapologetic embrace of commodification.

To represent such instrumentalization as 'co-optation' of a formerly radical and progressive postmodern architectural culture, as some commentators have suggested (Mills, 1988; Ley, 1989), is fundamentally wrongheaded. It postulates some original purity outside the economic nexus, but now debased and corrupted. Moreover, it hinges on a view that it is possible to decontaminate architectural and economic realities, separating the authentic postmodernist structure from its affirmative successors. But, if the firms working at Canary Wharf are at all indicative, such distinctions lose credibility. From the outset postmodern architects have accentuated the commercialization of the profession. Permeated by the language and ethos of product advertising (cf. Kieran, 1987; Bolton, 1988; Gutman, 1987), blithely resigned to the forces of commercial overdevelopment, convinced that architecture is so powerless as to be incapable of political significance (cf. Boyer in Chapter 5) and groping to maintain its own monopoly of competence, much of the architecture produced under the rubric of postmodernism has been blatantly commercial from its inception. In the work of KPF, Cesar Pelli or Phillip Johnson it sublimates commercial speculation to the highest form of the architect's art (Chao and Abramson, 1987). And, for more than a decade now, architects have at least matched their artistic pretensions with an explicit drive for financial success, dropping gentle professionalism to become 'hustlers' (Banham, 1982; Saint, 1983). The tangible outcome of this architectural boosterism is now written large in the exaltation and romanticization of corporate power at Canary Wharf and WFC.

Ideology and design

Moreover, urban design and architecture generate ideological effects insofar as they organize space and mobilize meaning in the built environment to sustain relations of domination. In the interrelation between statements of intention and the symbolism of design, we can locate the ideological significance of architecture and aesthetics at Canary Wharf and the WFC. On the one hand, in rationalizing their designs—a process aided by architectural commentary—architects generate narratives simultaneously revealing the assumptions grounding design and perpetuating mythological notions about urban transformation. As a means of legitimation, these narratives operate through various rhetorical strategies, only two of which are considered here. First, in the case of Canary Wharf, architect Cesar Pelli propagates what Holston (1989, p. 76) terms a 'foundation myth' in which he presents as naturally given, sacred and universal, a structure historically motivated. By invoking romantic precedents for his skyscraper at Canary Wharf, Pelli hopes to achieve a naturalization of the present by projecting it as an inexorable movement within history. Second, the permeation of architectural discourse by terms such as 'the public,' 'the community' and 'our times' sustains the belief that architects are not serving particular needs of profit but the general needs of the citizenry. Appeals to vague collectivities, whether in the form of *Zeitgeist* or communal social groups, have long been used as post-hoc justifications for stylistic choices of architects (Knox, 1987), but they also, especially in this case, serve to give redevelopment a semblance of democratic legitimacy.

On the other hand, if the city itself can be read as a text or considered as a discourse in which power is inscribed, then it behoves us to think about the message being projected *by* architecture. In a general sense, both Canary Wharf and WFC are spectacular diversions that draw a veil over the realities of deepening social polarization, ghettoization, informalization and burgeoning homelessness which characterize London and New York. As Harvey (1989b) suggests of the formulaic application of postmodern design tenets in inner-city 'regeneration' initiatives, they mimic a 'carnival mask that diverts and entertains, leaving the social problems that lie behind the mask unseen and uncared for. The formula smacks of a constructed fetishism, in which every aesthetic power of illusion and image is mobilized to mask the intensifying class, racial and ethnic polarization going on underneath.' This fetish gains its power through an aestheticism that postulates commitments to beauty and 'quality of design' as answers to social problems (Deutsche, 1988). For, within the terms of a simplistic critique of modernism and the liberation mystique of postmodernism, redevelopment is no longer oppressing people with urban 'renewal' and the blandness wrought by the international style but is offering an architectural 'renaissance' composed of 'meaningful' designs respecting their contexts and communicating with their users. In this sense, ethics are aestheticized—the good city becomes the beautiful and visually stimulating city and aesthetics becomes a weapon wielded by the forces of redevelopment, inserting social power more deeply into the experience and perception of everyday life in the city (see Lefebvre, 1971; Eagleton, 1990). But not just any aesthetic will attune consciousness to official representations of urban restructuring. At both Canary Wharf and WFC, the strategy is to engineer an aesthetic rooted in local historical conditions in order to familiarize the megastructure. The precise nature of these aesthetic ideologies can now be sketched.

Naturalizing speculation: The perils of aesthetic debate

Crucial to the legitimacy of O&Y's property ventures is acceptance in cultural and academic arenas. While the prolific aesthetic apologies for redevelopment formulated by influential figures such as Paul Goldberger, architectural commentator for the *New York Times*, have aided this in New York (cf. Deutsche, 1986), fierce controversy surrounds its intervention in London. Amid a hegemonic conservationist ethos, mounting controversy over architectural style (Hutchinson, 1989), an officially expressed Royal distaste for Canary Wharf's tower (cf. Charles, Prince of Wales, 1988), virulent attacks from the planning establishment (Tibbalds, 1989) and parochial distrust of the so-called 'invasion' of American practices, O&Y has been compelled to assume a more self-defensive posture. To this end, it utilizes cultural spokespersons as its vanguard. Thus, it has managed to command an institutional forum by dominating the Urban Design Group's annual memorial lecture for the past two years and enlisting Sir Roy Strong, former director of the Victoria and Albert Museum, as an advisor and cultural figurehead. Foremost among these efforts however are the rationalizations that Cesar Pelli offers for the skyscraper (Pelli, 1989).

Fundamental to his argument is a hierarchical distinction between 'true' skyscrapers and mere tall buildings. The latter are just vertical objects that do not carry the civic responsibilities of true skyscrapers. True skyscrapers, he maintains, are charged with representational responsibilities to act, by virtue of their towering height, as markers of place, sculptors of the city silhouette and as conveyors of public image. They must not finish abruptly in the flat top of modernist glass boxes, but culminate in a celebratory gesture. Working from such a premise he has explained his task in paternalistic terms as the design of London's first bona fide skyscraper. This is why the form is elementary and 'pure,' unmitigatingly vertical and undecorated. Clad in a shimmering stainless steel skin and crowned with a sleek pyramid, the tower is intended as an obelisk stripped of gothic or art deco trimmings that might link it too directly with American archetypes. As Pelli suggested in interview, he aspired to produce a form 'connected with history, but also beyond history' (*Architects Journal*, 1/11/89). When placed in the context of construction, this distinction loses it academic innocence. As a contemporary form of Pevsner's duality of 'building' and 'architecture,' it confers upon Canary Wharf a rarified aesthetic status, dissimulating its origins as a speculative property venture.

Proclaiming the timeless qualities of the building accomplishes this mystification even more thoroughly. In a revealing marketing statement (O&Y, 1990c) he disavows the historical contingency of redevelopment by appeal to a 'primeval' cosmic symbolism of humanity's urge to build ever upwards. Resorting to a crude form of biological functionalism all too common in architectural discourse, he justifies the tower through a narrative stressing the 'primordial,' 'constant' yearning of 'mankind' to build high structures. Though we cannot quite locate the 'psychological roots' of the urge he suggests, the historical record of tall structures is palpable evidence that it is nonetheless there, buried deep in the collective unconscious. As an adjunct to this, and in response to the negative reception of the tower in a London setting, a supplementary theme positions the building as just the latest incarnation of a centuries-old tradition in London to mark each age with a vertical expression (St. Paul's cathedral, Big Ben, Nelson's Column). The effect of this analogy is to dehistoricize the present and legitimate the tower in terms of the prestige and sanctity of these local icons. Like all 'foundation myths' Pelli's justifies the tower by recollecting precedents which he believes brought it into being and gave it is defining attributes. The precedent—'aspiring London' and the universal history of vertical structures—reveals not only a crude phallocentricity, but also invests the tower with natural and universal qualities, situating it as one of the most ancient, eternally valid components of urban structure.

A corollary of this mythological pedigree is to posit the tower's social site outside of history and beyond the 'normal' city. Anxious to convince that the tower does not destroy London's skyline, he argues the necessity of locating the tower in Docklands in the interest of townscape preservation:

Clearly, London is becoming a world financial centre and it has to accommodate

> *those functions, and I think one very healthy way to do this is to move a good distance to the periphery where towers can exist that will not affect or compete with St. Pauls or the Houses of Parliament or any of the other great spires, and also not change the scale of the city. London is a lovely city with a lovely small scale, smaller than Paris, smaller than Rome. That's why I think this peripheral development in a loosely attached satellite is the healthiest possible way.*
>
> Pelli, 1989

So, the underside to a postmodern concern for preserving the existing urban context is a negation of the socio-cultural context of east London. By portraying the Isle of Dogs as a 'periphery,' a mere appendage, Pelli not only gives graphic expression to the exploitative relationship that has historically obtained between the City of London and east London, but he also epitomizes, in terms closely fitting Lefebvre's concept of abstract space, the role of design professions in carving the city up into spaces of dominant centers and dominated peripheries. Indifferent to the social context into which Canary Wharf is inserted, he invents east London as a place of historical absence, a vast *tabula rasa* freely available for his experiments with London's first skyscraper. That there are many in the periphery who do not share his vision nor see the tower as a particularly advantageous way for London to 'grow' is effectively concealed. What matters according to this perspective is that London has an essential character, residing in its built structures, and that this is most effectively preserved by the siting the tower some distance from the center.

Undoubtedly, this argument is a coded response to Prince Charles' denunciation of the tower, but there is a fundamental affinity in the terms of debate. Certainly, there is a specific disagreement in that the Prince bemoans its disruptive impact:

> *Why does it have to be so high? In this country things have somehow always been on a more intimate scale . . . it's the size and density of Canary Wharf that has caused such a stir. It is the impact it will have on the London skyline—in particular on the famous view north from Greenwich—that is the problem.*
>
> Charles, Prince of Wales, 1988

But this should not be allowed to obstruct their common focal point: a formalistic concern with physical context and visual experience of the skyline. Attention focuses overwhelmingly on appearances and the question of the skyscraper's style in disregard of the social and economic issues which also contextualize the building. This is one instance of the unfortunate side of the Prince's intervention in architectural debate (cf. Wright, 1989). For it deflects attention away from social content and displaces it into crude polemics about visual detail, as if they were innocent of existing power relations, a separation exemplified in the appropriation of the City of London's monumental skyline as the referential for an authentic, populist London. In this evasion, discussions centering on height, visual axes, cornice lines, variations in color and shape and the whole scenography of the skyscraper aid and abet property developers. They confine criticism to the world of architecture and aesthetics, permitting the protagonists who wield financial and political power to fade into the background, their priorities remaining unexamined. As Ghirardo suggests, commenting on a parallel trend in architectural criticism in Houston, 'builders and developers could not in their wildest dreams have designed a strategy of such academic and intellectual status that it would successfully direct analysis toward trivial matters of surface and away from more vexing matters of substance' (1990, p. 236).

Representing the collectivity: constructing the public

This legitimation of Canary Wharf's skyscraper by historical analogy has a counterpart in Pelli's portrayal of WFC as a tangible expression of a harmonious culture. Though the terms do not match precisely, his comments recall Mies van der Rohe's famous assertion that architecture expresses the unifying spirit of the age:

> *I tried to design the buildings to be as intelligent and clear as a modern building but to also be responsible to the silhouette of the city, to the sidewalk, and the traditional buildings of Downtown New York. The whole issue to me was to do a building that is of our time, responding to our sensitivities, our ideology, the history that exists behind us, but without historic borrowings.*
>
> Cesar Pelli, quoted in Dean and Freeman, 1986

The notion that it is the architect's mission to intuitively discover the unified essence of an age and then fashion an appropriate built form has a long history in justifying stylistic choice and establishing the credibility of architects as custodians of 'the public interest' (Smith, 1971). As such, it is one of the institutionalized rhetorics mythologizing social conditions in order to sustain the discipline of architecture. In this case, the appeal to collective sentiments, like the repeated association of Canary Wharf with 'the community,' constructs an imaginary coherence which suppresses social difference and implies that dissenting visions are mere aberrations. By interpreting 'context' in narrow physical terms, Pelli disavows the deepening class and racial segregation characterizing New York's social context, preferring instead an idealist notion of society as an abstract, monolithic 'we.' His appeal to 'our' times and sensibilities is nothing more than a convenient fiction, a popular figure of speech among architects used to give form to a social reality that it does not transparently reflect, but constructs. Its significance is to imbue architecture with democratic credibility, since if architecture merely represents the life of 'the people' 'as passively as do the lens and plat of a camera' (Smith, 1971, p. 16), then it follows that the architect must be right by popular decree. From this standpoint, there is no need to ground design in a consideration of socio-economic context or urban politics; architects can simply retreat into metaphysics, exercising their supposedly preternatural sensitivity to the life and spirit of the age. Such homogenizing metaphors also predominate in the 'urban design' of Canary Wharf and WFC, expressed chiefly through the manipulation of 'public space' as a rhetorical pawn.

Urban design, because it claims the public realm as its key object of practice and theory, plays a crucial role in legitimating megastructures. Both projects are, in fact, acclaimed as generous additions to the public space of their respective cities, spaces that, so we are told, restore legibility and conviviality in the form of figural urban 'living rooms' (Goldberger, 1988). The triumph of public space is posited as a double victory. On the one hand it is deemed to rescue vital amenities from overdevelopment and to reverse an expensive, unscrupulous 'incentive zoning' system that reduces public space in New York to a residual melange of plazas and atria (cf. Lassar, 1990). On the other, both projects gain credibility by their conscious portrayal as repudiations of modernist planning's conception of space as continuous void and its anti-historicist approach to urban form. At Canary Wharf, for instance, the chief Landscape Designer (Lauri Olins) completely disavows modernism, and there is a definite effort to avoid the 'project look' of modernist urban renewal such as La Defense in Paris. In this attempt to disguise the origins of Canary Wharf as the biggest project of all, modernism is used as a foil to validate a restoration of traditional, familiar urban forms and typologies.

Paradoxically, since it is situated in an Enterprise Zone, Canary Wharf's development is governed by a rigid, axial urban structure and set of urban design guidelines mandating setbacks, the detailing of façades and the nature of materials. According to O&Y's 'master' planner, these provide prior structuring elements, ensuring that architects do not produce 'rootless postulations of new urban forms' which conflict with their desire to produce a 'normal city' (Coombes, 1990). Beyond its role in adding value, therefore, the 'release' of one-third of Canary Wharf's 72-acre site for public use is intended to provide aesthetic coherence and establish the social physiognomy of the project, thus replicating the function that the infamous plaza played in the development of Rockefeller Center (Balfour, 1978). Morphologically, Canary Wharf is a bricolage of perspective axes, spectacular civic squares, monumental entry points and urban parks, all unified as part of a revival of a 'grand tradition' (Strong, 1988) which is considered 'universally acceptable' because '4000 years of urban history can't be wrong' (Coombes, 1990). At a general level, by reviving the city square, mandating street walls clad in crafted masonry and installing room-like streets at the heart of the megastructure, the designers of Canary Wharf were deliberately using forms and conventions that have long constituted and signified the public domain. Their strategy was to appropriate an architectural order carrying historically accrued connotations of democracy, citizenship and public congregation, thus yielding instant civic credentials (cf. Jencks and Valentine, 1987; Rowe and Koetter, 1978).

Additionally however, these forms are given a local inflection. In a maneuver that corresponds directly to Rossi's (1982) advocacy of urban typologies embedded in the 'collective memory' of the city's inhabitants, the designers of Canary Wharf have consciously tried to discover what makes London's public sphere memorable and then incarnate it in new, larger forms. This is metonymical in that it does not literally reinvent the architecture of London's past, but rather, by evoking the feel and shape of popular outdoor spaces, seeks to stimulate the sense of place and history associated with them. Hence the typologies, street furnishings and façades all deliberately play upon memory, metaphor and symbolism in a calculated strategy to condense all the richness, texture and diversity of London into a single space. This strategy also effectively negates the urban order of modernism, familiarizing the megastructure in an attempt to assimilate it into the everyday perception of London. Each of its main spaces is derivative from this repository of local historical forms.

At the western entry to Canary Wharf stands West Ferry Circus, a decorous homage to London's tradition of parks. Modelled on St James's Park and Finsbury Circus in the City of London, it is intended as a soft landscape conducive to passive relaxation, a resurrection of the *rus in urbe* tradition (Strong, 1989). This leads into West India Avenue, a 'broad and noble avenue' based on the axial convention of baroque and City Beautiful planning. A grand setpiece in the manner of London's Mall, it aspires to capture its imperialist connotations of monumentality. By its alignment with vistas to the City of London and its eastern culmination in Pelli's landmark tower, this axis sets up a visual structure connecting these two nodes of power, communicating unambiguously the prestige of O&Y. For this purpose, the baroque model is appropriate, since it has historically always been closely linked with elites. From the rebuilding of Paris under the aegis of Haussmann, this baroque form has historically worked as a 'strategy of awe' (Lasswell, 1979), manipulating the city as an expression of central power through visual magnificence. West India Avenue eventually opens out into Cabot Square, a hard, formal rendition of Trafalagar Square complete with sculpture and fountains. Even the architectural detailing in this square is calculated to evoke analogies with 'Old London,' as Adrian Smith, architect for SOM, reveals of the building he designed:

> *No. 3 Cabot Square recalls the spirit, human scale, and texture of the traditional buildings of London. Rather than referring specifically to any one building, architect, style or historical period, it captures and distills the essence of the entire region.*
>
> (quoted in O&Y, 1990d)

Opulent decoration of these spaces carries the strategy through to the level of fine detail. It also discloses the Corporation's intent to thoroughly aestheticize everyday life by painstaking attention to detail. Every visual and sensory delight is catered to in an assortment of custom-designed fittings commissioned from the world's finest craftspersons. Thus here will be: 2 km of waterfront walkways and tree-lined boulevards; richly textured glass fittings; manicured gardens and mature vegetation imported from Europe; exhilarating watercourts with stunning waterside vistas; elegant fountains moving through colorful displays; patented street lighting; Art Nouveau park railings designed by Buiseppe Lund; and solid teak benches crafted by Rod Wales. Such attention to sensory pleasures is not just about producing a competitive business environment, it is also proclaimed as a creative reinvigoration of the public realm. It is a 'strategy of admiration' (Lasswell, 1979) bidding for acceptance through histrionic displays of beauty, cultivation of the picturesque and evocation of nostalgia.

The design philosophy adopted at WFC replicates this. As part of a contextualist design framework aiming to create Battery Park City as a facsimile of older New York neighborhoods, the towers of WFC also try to integrate with downtown Manhattan. Their setbacks, sculpted tops and graduated mix of granite and glass establish visual and perceptual connections with the surrounding city, consciously echoing the romanticism of interwar skyscrapers. Again, detailed design of public spaces reinforces this effect. Fragmentary reminders of New York's public life are selectively incorporated into the plaza and Winter Garden. The latter, for example, besides containing 16 palm trees, is rhapsodically claimed by its architect to be 'everybody's living room' and its grand semicircular stairway is planned as a 'handout' to match the casual use of the New York Metropolitan Museum of Art's entrance (Pelli, 1990). The surrounding lobbies too are reminiscent of the sumptuous interiors of New York's earliest, popularly cherished skyscrapers such as the Woolworth building. Outside, the plaza, enthusiastically billed as an artwork in itself, mixing 'poetry and public service,' is decorated with signs of locality. Cast-iron railings crafted by artist Siah Armajani carry poetic commemorations of old New York drawn from Walt Whitman and Frank O'Hara. Classic New York City park furniture is retrieved to furnish its verdant gardens. And even granite benches produced by sculptor Scott Burton aim to produce an art form not only socially useful, but evocative. As he explains:

BARNEYS NEW YORK THE HUDSON RIVER AU BON PAIN THE ESPLANADE LOIS LANE TRAVEL FOOT AUTHORITY OPTOMETRIC ARTS DONALD SACKS COBBLER EXPRESS ST. MORITZ CHOCOLATIER UNITS PAPER BOUTIQUE EASTERN LOBBY SHOPS BREE LEGS BEAUTIFUL PLUS ONE FITNESS CLINIC TWIGS CASWELL-MASSEY THE STATUE OF LIBERTY

City Tales :

DESTINATIONS

Arrivals & Departures

by DANA GIOIA

photograph by Elizabeth Zeschin

Today most travel feels like commuting. Distant airline journeys begin and end by waiting in line. Cars find the traffic in exotic cities drearily familiar. The same bored faces look up from bus station benches from Antwerp to Alabama. There is no poetry in reaching a public garage, a bus stop or luggage carrousel. Just the flat, minimal prose of baggage claims and parking meters. ❧ But there is one commute that actually feels like travel –coming into Grand Central Terminal. Entering its vast main lobby either by descending from the noisy street or by rising from the steamy train platforms underground, one walks into Manhattan's greatest indoor public space, an area all the more exciting because unlike its closest contenders – Carnegie Hall, the Stock Exchange, the Palm Court at the Plaza – Grand Central is neither elitist nor exclusionary. In its democratic precincts investment bankers in snug italian suits line up for coffee behind Hawaiian-shirted street vendors and elderly nuns on a shopping spree share waiting room benches with honeymooners from Osaka. ❧ Stepping onto Grand Central's smooth marble floor is always like walking onstage but whether into a stark social drama or screwball screen comedy one can never quite tell in advance. Hundreds of lives hurry by, each one starring in its own compelling story. Charged by their sheer energy, one feels that particular rush of excitement that only great cities give. This almost delirious feeling that anything is possible represents the unacknowledged triumph of Grand Central's architecture. Utterly functional, this station is also truly grand. Its one vast central chamber stands impressively surrounded by teeming platforms, passageways, tunnels, balustrades, and antechambers, the air above crisscrossed by huge shafts of filtered sunlight rising to the high arched ceiling decorated astonishingly –as if to say that this one room is indeed its own universe– by the stars and constellations of the zodiac. Yet even to a gaping first-time visitor, a lone pedestrian clutching a battered suitcase on the swirling concourse below, all this grandeur seems not only unintimidating but inviting. This is a place which recognizes the importance of each arrival. Nervous, giddy, even inspired, one steps into the crowd ready to begin.

Over 30 thousand people come to work or play at The World Financial Center every day. They come by cars, boats, trains and planes. Once they come, they stay till all hours. Come and see what all the fuss is about.

KODO DRUMMERS RECTOR PLACE NEW YORK NEW YORK IT'S MORE THE WINTER GARDEN LIBERTY STREET AT BATTERY PARK CITY

PERL PHOTO ELECTRONICS GALLERY OF HISTORY MANUFACTURERS HANOVER TRUST STATE OF THE ART MINTERS IMMAGA VESEY STREET ANN TAYLOR SIXTEEN PALM TREES MARK CROSS RIZZOLI

Fig. 6.3. 'Arrivals and Departures.' One of a series of advertisements for the World Financial Center. Reproduced courtesy of Drenttel Doyle Partners, New York.

> *In the same way that the tops of the buildings are designed to echo skylines the esplanade is designed to echo New York City parks, the sculptures—on totally contemporary bases and installations—will provide some continuity with old New York. We want to reconnect with what survives and what is familiar.*
>
> (quoted in Trilling, 1985)

Such intentions correspond with the format behind an early promotional campaign for the Center, illustrating a convergence in the meanings produced by architects, public artists and advertisers (Fig. 6.3). Titled 'City Tales' and prefaced as 'It's more New York New York,' it consisted of a series of advertisements initiated in October 1988 aiming to stimulate awareness of the Center's newly opened public spaces and upscale retail gallery. Seemingly unconnected to the WFC, the advertisements involved short literary vignettes and black and white photographs focusing on icons of New York public life such as Grand Central Terminal and Korean fruit stands. Their subtle rhetoric intended to tap people's warm sentiments about New York and then transfer them to WFC, aided by typographics which literally frame New York within its precincts. They strive to generate an immediate sense of belonging for the Center in the weekly round of up-market consumers—an effect accentuated by their placement in publications such as *Vanity Fair*, *Time* and *The New Yorker* and the mysterious format of their message.

Public space, particularly when it is designed beautifully and in accord with a supposed consensus on the virtue of historicism, has therefore become a token of commitment to meeting public needs. It is simplistically portrayed as the vibrant social space of a harmonious community. Moreover, in the absence of prominent public buildings and public-sector commissions in either Battery Park City or Docklands, O&Y can claim to be the key saviors of the public realm. 'Public space' at the WFC and Canary Wharf is also held to be emblematic of citywide vitality. In the propaganda of City boosterism, nothing signifies ascendancy and livability more powerfully than the idyllic picture of 'the public' strolling, relaxing and soaking up the ambience of these opulent spaces. Yet these representations impose several mystifications in the way they construct and manipulate the concept of 'the public.'

For one thing, the rhetoric of public space masks the nature of the processes producing the megastructure. Just when O&Y is truly demonstrating the capability to level local difference through the abstractions of a globalizing capitalist property industry, it provides superficially reassuring images of place and promises the restoration of a harmonious, ideal urban environment rooted in tradition. As Canary Wharf obliterates remaining traces of a local Docklands history, for example, its re-constituted borrowings present public space as a cinerama architectural tour of London, offering an experience of a place more perfect and authentic than the original. A gigantic piece of pastiche, it seeks 'our' habituation to its version of the future by evoking 'our' imaginary past and using symbolic codes already intimately known to us. The very notion of the public sphere at work—public space as physical volumes, public art as any art situated outside of the gallery or museum—conspires to hide from us the widespread privatization of the public realm and its reduction to the status of commodity. It disguises, for instance, the fact that, in the case of Canary Wharf, such space is constructed on public land appropriated for private interests, is fashioned by a process insulated from public input and will be controlled, not by a democratically elected local authority, but according to the logic of O&Y's management.

Under this logic, public space conceived as an arena of democratic political debate is destroyed in favor of a public realm deliberately shaped as urban theater. Significantly, however, it is theater in which a pacified public basks in the grandeur of a carefully orchestrated corporate spectacle. Through symbolic means (fortress-like entrances and walkways) and elaborate surveillance techniques (from a paid security force to a panoptic video system), access to the WFC is highly selective, producing an anesthetized social world depolluted of antagonism and social conflict. Once refined, the resulting public is there for entertainment and there to gaze. Viewed this way the Winter Garden is less a forging ground in which collective experience is given group expression and more a space of private serenity acting as a glamorous showcase for the model public of the late capitalist city. The architectural semiotics of these spaces signal how this public should function. The only 'active' experiences clearly sanctioned are those of conspicuous consumption and exhibitionism, as the architect candidly suggests:

> *It is one of the most wonderful qualities of urban life, for us all to be actors and spectators. We have designed the Winter Garden very carefully so that this pageantry is continuously active and reinforcing itself . . . when people walk up that stair, they know they are being watched, and they walk a little more erect. They feel handsome; they know they are on-stage and it is a wonderful experience.*
>
> Pelli, 1990

All other activity is to be 'passive': rest, contemplation, quiet consumption of expensive cuisine and admiration of the picturesque maritime scene in the plaza. Likewise, Canary Wharf is not designed to allow for the outdoor activities and political demonstrations that are part of the history of the social use of London's public spaces. It too offers the fantasy of community and cohesion stimulated through aesthetic experience and a common gaze upon comforting, escapist spaces. Such fantasy only persists through banishing unsightly realities to places elsewhere. Ultimately, the public at the megastructure is constructed as a mass of spectators and consumers who will visit it as a 'destination,' akin to landmarks on a tourist heritage trail.

Further, contrary to the architect's best wishes, the Winter Garden hardly functions as an accessible 'living room' for New York. This illusion of a homogeneous public depends upon exlusionary procedures which repress competing social realities disrupting the image of the megastructure as an egalitarian benefit for all citizens. Spaces such as WFC are 'programmed' to filter the social heterogeneity of the urban crowd, substituting in its place a flawless fabric of white middle class work, play and consumption—'bourgeois playgrounds' in Smith's (1987) words—with minimal exposure to the horrifying level of homelessness and racialized poverty that characterises New York's street environment. This dissociation of the space of the privileged from the space of the dispossessed is accomplished physically and discursively. Physically, segregation is carefully maintained by a whole apparatus of repressive spatial tactics: through ruthless eviction of the homeless from public spaces and their confinement to officially designated shelters, safely out of sight; by eliminating public amenities that might make city streets minimally 'livable'; by sequestering formerly public space for private use through construction of 'defensible spaces'; and less invidiously, but scarcely less effectively, through linkage agreements promising 'off-site' benefits in the form of a grossly inadequate number of low income housing units, which in reality only entrench the ghettoization of low-income residents in Harlem and the South Bronx.

Bolstering such forceful means, severance is attained discursively by a whole series of propagandist conceits externalizing poverty. It is accomplished, for example, by neutralizing homelessness as an inevitable condition; by representing homeless people as social deviants responsible for their own condition quite apart from inadequacies of housing provision; and even by constructing entire inner-city districts as discrete entities beyond the normal city (Keith and Rogers, 1991). Whether such representations originate in parliamentary reports, municipal documents or advertisements, their effect is to 'restore to the city a surface calm that belies underlying contradictions' and to expel certain social groups from representation of the city (Deutsche, 1988, p. 5). It is precisely this appearance of calmness that legitimates WFC and Canary Wharf as 'public' spaces. The activation of these spaces by art contributes to this illusion of publicness.

Marketing real estate with art

For advertising purposes, I felt that Matisse and Picasso has the greatest drawing value . . . while as for Rivera, although I do not personally care for much of his work, he seems to have become very popular just now and will probably be a good drawing card.

Nelson Rockefeller, quoted in Balfour, 1978, p. 132

Though concerned with the arts program for his own mega-development of the 1930s, Rockefeller's intent approximates to that generating the current surge of corporate investment in culture (cf. Scardino, 1987; Zukin, 1991). Matching the attempts by city governments to build a soft infrastructure appropriate to consolidating New York and London as world cities, individual developers also compete culturally, seeking to enhance their competitiveness by providing alluring packages of public art, lunchtime performances and expensive art collections. Thus, with Broadgates 'Circle' providing a performance amphitheatre and its developer investing lavishly in sculpture, Canary Wharf could not afford to be without its kit of cultural parts and had to emphasize its devotion to the finest details of the so-called 'people spaces' between buildings. Yet beyond the anchorage it provides for property values and its powerful role in generating tenant loyalty, art also bolsters public credibility. It testifies to the cultural discernment of the possessor and establishes the 'distinction' of redeveloped spaces because it is perceived as disinterested and presumed to transcend politics (Bourdieu, 1984). If expenditure on public art is one dimension of this attempt to refashion the urban 'wasteland' into a pleasurable paradise, then the aestheticization of everyday life is completed by the production of spectacles designed to bring culture to 'the people.'

This additional cultural cladding is provided by O&Y's patronage of Arts and Events Programs to 'animate' public spaces. As integral parts of their 'total environments,' O&Y incorporate structures designed to operate as urban theatres, loosely modelled on the Greek agora. Thus, the Winter Garden converts into an 'informal' cultural venue and Canary Wharf will include a permanent concert hall. A team of arts managers coordinates privately funded free-to-the-public performing and visual arts events in these spaces, offering 'civilized metropolitan fun' and transforming the megastructure into a display case for both established and emerging artists. The program at WFC, for instance, previews programs scheduled elsewhere in New York and commissions its own work between three and five times per year. Although the 30,000 workers at the Center form the primary audience, performances have managed to attract a statewide crowd of between one and two thousand at each event. The most spectacular event staged so far was a 'New Urban Landscapes' exhibition celebrating the opening of the WFC's public spaces in October 1988 (Martin, 1990). By commissioning 28 installations that attracted 50,000 visitors, O&Y claimed to be inviting a public 'dialogue' about the urban environment. The organization of this exhibition reveals, however, the ability of the corporation to dictate the terms of debate.

The twenty-eight participating artists and architects in the exhibition addressed a variety of issues relevant to contemporary urban experience: from societal attitudes towards shelter to the changing nature of recreation; from the powers of technology to the importance of media representations in the city. Presented as socially and politically engaged, its radicality exacerbated by its command of New York's pre-eminent space of richness, the 'New Urban Landscape' was congratulated as a 'critical' confirmation of 'the unmitigated truth of art' (Martin, 1990, p. 20). Yet this adulation conceals the exclusion of art forms that proved too challenging to the Corporate perspective on urban change: architectural contributions which conflicted with Pelli's modernism and artistic projects whose social critique was too powerful, though solicited, were eventually rescinded.[8] And, none of the exhibits seriously questioned their institutional context, drawing short of a self-reflexivity that would illuminate the role played by specific cultural institutions in the process of redevelopment. Instead, the art illustrates urban life in a depoliticized form. Even the most disquieting installations, such as Tadashi Kawamata's 'Favela in Battery Park City: inside/outside'—a crude lean-to structure contrasting with the permanent opulence of corporate architecture—leave their urban commentaries vague and enigmatic, never daring to expose the specific connections between the redevelopment and homelessness in New York. So, though this art may not be politely reconciled with its environment, its overall context conditions its reception far more than the particularity of its messages. Ultimately, the exhibition was part of a staged urban spectacle and one more instance of aesthetic amenities as advertising. This was unequivocally signified by the appropriation of one of the exhibits as an advertisement for the show which has since become the logo for the WFC's promotional literature. The exhibition effectively permitted the corporation to display its assumption of civic responsibilities that might once have been undertaken by public institutions and city governments. O&Y's strategy is, in fact, the latest permutation in the relationship between art institutions and redevelopment. For, instead of sponsoring exhibitions in existing institutions or providing funds to maintain existing galleries, O&Y seems determined to posit the corporate environment itself as a cultural arena: a new art institution. This is itself a product of the search for ever more novel ways to show a largesse of public concern over and above that evidenced by competing developers.

Significantly, however, this boom in corporate art commissioning proceeds in step with the decimation of cultural activity in most of the inner city. As a recent arts action program from the London Docklands Development Corporation contradictorily acknowledges, the systematic attempts by developers to buy cultural furnishings 'off the rack' has its underside in the reduction of public funding for the arts, the displacement of existing arts activities in east London, the closure of local theaters and a struggle to survive dismantling of all cultural institutions seen as critical of the right-wing political agenda (LDDD, 1990). In keeping with the logic of redevelopment, therefore, cultural

[8] Both Kryzsztof Wodiczko (artist of the homeless vehicle project) and Ben Nicholson (who submitted a piece of deconstructionist architecture overtly challenging modernism) both had proposals subsequently rejected, ostensibly on grounds of safety (see Schwartzman, 1989; Lurie and Wodiczko, 1988).

provision is drastically uneven: concentration of affluence in official corporate centers and impoverishment of alternative cultural forms in the spaces where developer self-interest does not extend.

Concluding comments

The construction and presentation of the megastructure as a nostalgic paradise regained thus marks a distinctive moment in the form that urban restructuring takes. Aesthetically, it differentiates these developments from the austerity and minimalism of modernist urban 'renewal' and also from the reflective-mirror surfaces of contemporaries such as the much discussed Bonaventura Hotel in Los Angeles (Jameson, 1984). They no longer attempt to defamiliarize and shock as the protagonists of modernism intended with their use of new conventions, materials and architectural 'vocabularies' but instead attempt to accommodate by using an architectural language systematically learned from London and New York. Likewise, instead of the placeless dissociation and repulsion of the outside city that Bonaventure achieves, Canary Wharf and WFC purport to recapture the spirit of enchantment fostered at Rockefeller Center, proffering the charms and delights of the traditional city. Yet this return of enchantment, itself an embodiment of the ideology of postmodern architecture, should not be accepted as benignly as liberal commentators such as Ley (1989) aver. As Tafuri (1980) argued in relation to Rockefeller Center and Boyer (1988), more recently in relation to the wholesale return of aesthetic priorities to city planning, aestheticization of the speculative functions of real estate to this extent dissembles more serious grounds for *dis*enchantment about contemporary urban change.

The populist aspirations of the megastructure to function as a microcosm of the rest of the city, for instance, can be read negatively as an extreme instance of the bureaucratization of everyday life. Traditional cities, with their connotations of vitality, social interaction and heterogeneity, cannot be 'programmed' or 'animated'; history and memory in the city do not have 'essences' reducible to visual images; and a genuine public presence cannot be engineered through application of correct forms, dazzling spectacle or the lure of free bread and circuses. Similarly, a meaningful city does not equate with those meanings packaged, marketed and prescribed by postmodern stylists. It arises from conflicts over the use of the city and a process of attribution by the users of the city. To be sure, Canary Wharf seems destined to match the popularity of its Manhattan counterpart, but this should not be used to disguise the instrumentalism behind its consumerism, nor should it lead us so quickly to postulate crowd practices in such spaces as effective modes of resistance, as Shields (1989) advocates in his consideration of shopping malls. There is a world of difference between minor transgressions of prescribed dress and behavior codes in these spaces and those larger acts of spatial subversion in which the public has periodically reappropriated places of power for demonstrations of solidarity and oppositional practice. Individualized rebellion in the consumption spaces of the megastructure still leaves us a long way from those collective acts of spatial reclamation involved, for instance, in the Paris Commune (Ross, 1988), the occupation of New York's Tompkins Square Park in 1988 as an anti-gentrification demonstration (Smith, 1988) or the conquest of London's Trafalgar Square in 1990 by anti–Poll Tax demonstrators.

The seeming implausibility of such a practical strategy is exacerbated by the closures enacted by aesthetic debate. The more that the megastructure can be discussed in the supposedly neutral, depoliticized categories of aesthetics, then the more elitist and detached from popular debate about the future of the city it becomes. The prioritization of aesthetics makes it even more difficult for locally-based opposition groups to contest a development already backed by a coalition of nationally and internationally constituted actors. It renders resistance doubly more difficult, since, on the one side, it makes protest seem like a rejection of modernity and civic benefits and, on the other, it has the effect of marginalizing the distributional questions which contextualize the megastructure. To restore questions of homelessness, housing crises and unemployment to discussions of urban restructuring, it has now become necessary to break through the clouds of aesthetic debate. It is salutary, for instance, that the articulation of these issues in the debate surrounding Canary Wharf has been left to those local protest groups with least formally empowered voices (cf. Brownhill, 1990; DCC, 1990). Against the flagrant disregard for much-needed public housing projects, they have been left to emphasize the opportunity costs Canary Wharf imposes upon such a project through its appropriation of land and public finance. Against a rhetoric of social integration and working for the community, they have stressed the subordinate, dependent terms under which this economic integration will proceed. Against the planners' outrage at the inhumane size and scale of the venture and the violation of venerable 'English' town-planning traditions it inflicts, they have been a lone, persistent voice stressing the inefficiency and inequity of planning London as a collage of 'projects.' And, against acquiescent fascination in the project as the latest spectacle of urban modernity, critical forms of public art have visualized local groups projections for alternative futures. Nor should this be maligned as a sentimental localism that pugnaciously asserts the counter-hegemonic. The issues involved are of international urban significance as a comparative analysis with New York would indicate. There, the fact that popular opposition to Battery Park City has been far more mute in the absence of a tenacious working-class presence in Downtown districts does not obviate the need to illuminate and contest the connections between WFC and deepening social polarization. Nor is it sufficient to allow architects and designers who willingly marginalized low-income housing from these projects to present such issues as 'beyond' their sphere of influence. After all, it is such retreats into defensive professionalism and artificial separations of the aesthetic from the sociopolitical which makes architecture and current forms of urban design such useful stakes in capital's continuing subordination of urban space to the logic of profit.

References

Balfour, J. 1978 *Rockefeller Center: Architecture as theatre.* New York: McGraw-Hill Book Company.

Banham, R. 1982 'The architect as gentleman and the architect as hustler,' *RIBA Transactions* 1, 33–38.

Blake, P. 1976 *Form Follows Fiasco.* Boston: Little Brown.

Bolton, R. 1988 'Architecture and Cognac' in J. Thackara (ed.), *Design After Modernism.* London: Thames and Hudson, 85–95.

Bourdieu, P. 1984 *Distinction: A Social Critique of the Judgement of Taste.* London: Routledge and Keegan Paul.

Boyer, M.C. 1988 'The return of aesthetics to city planning,' *Society*, 25 (4) 49–56.

Brain, D. 1989 'Discipline and Style: the Ecole des Beaux Arts and the social production of an American Architecture,' *Theory and Society*, 18, 807–68.

Brindley, R., Rydin, Y., and Stoker, G. 1989 *Remaking Planning.* London: Unwin Hyman.

Brownhill, S. 1990 *Developing London's Docklands: Another Great Planning Disaster?* London: Paul Chapman Publishing Ltd.

Burgee, J. 1987 'The value of good building design . . . and the myths that keep us from it,' *Urban Land*, August, 12–15.

Chao, S.R. and Abramson, T.D. 1987 *Kohn Pedersen Fox: buildings and projects 1976–1986.* New York: Rizzoli.

Charles, Prince of Wales. 1988 'A Vission of Britain,' *BBC TV transmission*, October 28.

Church, A. 1988 'Urban regeneration in London Docklands: A five year policy review,' *Environment and Planning C: Government and Policy*, 6, 187–208.

Church, A. 1990 'Waterfront regeneration and transport problems in London Docklands,' paper presented to the Institute of British Geographers, annual conference, Glasgow.

Coombes, A. 1990 Annual Kevin Lynch Memorial lecture, reported in *Architects Journal*, 4 July 1990, p. 15.

Cox, K. and Mair, A. 1988 'Locality and community in the politics of local economic development.' *Annals of the Association of American Geographers* 78, 307–325.

Crilley, D. 1990a 'Image and counter-image: Contrasting meanings of redevelopment in London's Docklands,' paper presented to the Association of American Geographers, annual conference, April 21, Toronto.

Crilley, D. 1990b The Social Production Postmodern Architecture. Unpublished manuscript, Department of Geography, Queen Mary and Westfield College, London E1 4NS.

Crimp, D. 1989 'Hot in the 1980's: the myth of autonomy,' *PRECIS*, 6, 83–91.

Davis, M. 1988 'Urban Renaissance and the Spirit of Postmodernism' in E. Ann Kaplan (ed.), *Postmodernism and its Discontents*, London: Verso, 79–88.

Dean, A. O. and Freeman, A. 1986 'The Rockefeller Center of the '80s?' *Architecture*, December 1986.

Deutsche, R. 1986 'Krystof Wodiczko's homeless projection and the site of 'urban revitalization,'' *October 38* (Fall), 63–99.

Deutsche, R. 1988 'Uneven development: Public art in New York City,' *October 47* (Winter), 3–53.

Docklands Consultative Committee. 1988 *Six Year Review of the LDDC*, DCC, Unit, 4 Stratford Office Village, Romford Road, London E15.

Docklands Consultative Committee. 1990 *The Docklands Experiment: A critical review of eight years of the London Docklands Development Corporation*, Unit 4, Stratford Office Village, 4 Romford Road, London E15.

Duffy F. and Henney, A. 1989 *The Changing City*. London: Bulstrode Press.

Eagleton, T. 1990 *The Ideology of the Aesthetic*. Oxford: Basil Blackwell.

Feagin, J.R. 1988 *Free Enterprise City: Houston in Political-Economic Perspective*. New Brunswick: Rutgers University Press.

Foster, H. 1985 *Recodings: Art, Spectacle, Cultural Politics*. Port Townsend, Washington: Bay Press.

Gill, B. 1990 'Battery Park City,' *New Yorker*. August 20, 69–78.

Ghirardo, D. 1987 'A taste of money: Architecture and criticism in Houston,' *Harvard Architecture Review*, 6, 87–97.

Ghirardo, D. 1990 'The Deceit of Postmodern Architecture' in Shapiro, G. (ed.), *After the Future: Postmodern Times and Places*. Albany: State University of New York Press, pp. 231–255.

Goldberger, P. 1986 'Battery Park City is a Triumph of Urban Design,' *New York Times*, August 31, p. H1.

Goldberger, P. 1988 'Winter Garden at Battery Park City,' *New York Times*, October 12, p. C15.

Gottlieb, M. 1985 'Battery Project Reflects Changing City Priorities,' *New York Times*, October 18, P. B1.

Gutman, R. 1987 *Architectural Practice: A Critical Review.* Princeton: Princeton Architectural Press.

Haila, A. 1988 'Land as a financial asset: The theory of urban rent as a mirror of economic transformation,' *Antipode*, 20, pp. 79–101.

Harris, P. et al. 1982 'The marketing of meaning: Aesthetics incorporation,' *Environment and Planning B: Planning and Design*, 9, 457–66.

Harvey, D. 1989a 'Downtowns,' *Marxism Today*, January, p. 21.

Harvey, D. 1989b *The Condition of Postmodernity*. Oxford: Basil Blackwell.

Harvey, D. 1989c 'From Managerialism to Entrepreneurialism: The transformation in urban governance in late capitalism,' *Geografiska Annaler* 71(B), 3–17.

Holston, J. 1989 *The Modernist City.* Chicago: University of Chicago Press.

Hutchinson, M. 1989 *The Prince of Wales: Right or Wrong? An Architect Replies.* London: Faber and Faber.

Huxtable, A. L. 1984 *The Tall Building Artistically Reconsidered: The Search for a Skyscraper Style.* New York: Pantheon Books.

Hylton, R. 1990 'Olympia and York Selling Stake in U. S. Holdings,' *New York Times*, September 20, p. D1.

Jameson, F. 1984 'Postmodernism or the cultural logic of late capitalism,' *New Left Review*, 146, 53–92.

Jencks, C. 1987 *The Language of Post-modern Architecture*, 5th edition. London: Academy Editions.

Jencks, C. 1988 *Architecture Today*. London: Academy Editions.

Jencks, C. and Valentine, N. 1987 'The architecture of democracy,' *Architectural Design*, vol. 57, No. 9/10, 7–28.

Keith, M. and Rogers, A. 1991 *Hollow Promises: Policy and Practice in the Inner City*. London: Mansell Press.

Kieran, S. 1987 'The architecture of plenty: Theory and design in the marketing age,' *Harvard Architecture Review*, 6, 102–13.

King, A.D. 1990 *Global Cities: Post-imperialism and the Internationalism of London.* London: Routledge.

King, R. J. 1988 'Urban design in capitalist society,' *Society and Space, Environment and Planning D*, vol. 6, 445–74.

Knox, P. 1987 'The social production of the built environment: Architects, architecture and the postmodern city,' *Progress in Human Geography*, 11, 354–77.

Koolhaas, R. 1978 *Delirious New York*. London: Academy Editions.

Lassar, T. J. 1990 'The limits of incentive zoning,' *Urban Land*, May, 12–15.

Lasswell, H. 1979 *The Signature of Power*. New York: Transaction Books.

Lefebvre, H. 1971 *Everyday Life in the Modern World*. New Brunswick: Transaction Publishers.

Lefebvre, H. 1979 'Space: Social Product and Use Value' in J. W. Freiber (ed.), *Critical Sociology: European Perspectives*. New York, Irvington Publishers, 285–95.

Ley, D. 1989 'Modernism, Postmodernism and the Struggle for Place' in J. Agnew and J. S. Duncan (eds.), *The Power of Place*. London: Unwin Hyman, 44–65.

London Docklands Development Corporation. 1990 *Creating a Real City: An Arts Action Programme For the London Docklands*. LDDC, Great Eastern Enterprise, Millharbour, London E14.

Lurie, D. and Wodiczko, K. 1988 'Homeless vehicle project,' *October 47* (Winter), 53–68.

Marder, T. A. (ed.) 1985 *The Critical Edge: Controversy in Recent American Architecture*. Cambridge, MA: MIT Press.

Martin, R. (ed.) 1990 *The New Urban Landscape*. New York: Olympia and York Companies and Drenttel Doyle Partners.

Mills, C. A. 1988 "'Life on the upslope": The postmodern landscape of gentrification,' *Society and Space: Environment and Planning D*, vol. 6, 169–89.

National Audit Office. 1988 *Department of the Environment: Urban Development Corporations*, Report by the Comptroller and Auditor General, HMSO, London.

Olympia and York. 1990a *Canary Wharf: The Untold Story*. 10 Great George Street, London, SWIP 3AE.

Olympia and York. 1990b *Canary Wharf: Vision of a New City District*. 10 Great George Street, London, SWIP 3AE.

Olympia and York. 1990c *The Tower No. 1 Canada Square*. 10 Great George Street, London, SWIP 3AE.

Olympia and York. 1990d *Ogilvy & Mather No. 3 Cabot Square Canary Wharf*. London E 14. Olympia and York, 10 Great George Street, London SWIP 3AE.

Pelli, C. 1989 'The mega building on context,' *Architectural Design*, vol. 58, no. 11/12, 40–44.

Pelli, C. 1990 'Cesar Pelli' (interview) *World Financial Center News* No. 1, Olympia and York Companies, 200 Liberty Street, New York, NY 10281.

Rabenek, A. 1990 'Broadgate and the beaux arts,' *Architects' Journal*, vol. 192, no. 17, October 24th, 36–57.

Robson, D. 1990 'Canary Wharf,' *The Planner*, October 12, 8–10.

Ross, K. 1988 *The Emergence of Social Space: Rimbaud and the Paris Commune.* London: Macmillan.

Rossi, A. 1982 *The Architecture of the City.* Cambridge, MA: MIT Press.

Rowe, C. and Koetter, A. 1978 *Collage City.* Cambridge, MA: MIT Press.

Saint, A. 1983 *The Image of the Architect.* New Haven: Yale University Press.

Savills. 1990 *Central London Office Demand Survey*, 20 Grosvenor Hill, London W1.

Scardino, A. 1987 'Marketing Real Estate with Art,' *New York Times*, February 9, p. C23.

Schwartzman, A. 1989 'Public art—corporate trophies,' *Art in America*, February.

Shields, R. 1989 'Social spatialisation and the built environment: The West Edmonton Mall,' *Environment and Planning D: Society and Space*, 7, 147–64.

Smith, A. 1989 'Gentrification and the spatial constitution of the state: The restructuring of London's Docklands,' *Antipode* 21, 232–60.

Smith, N. 1984 *Uneven Development: Nature, Capital and the Production of Space.* Oxford: Basil Blackwell.

Smith, N. 1987 'Of yuppies and housing: gentrification, social restructuring and the urban dream,' *Environment and Planning D: Society and Space*, 5, 151–72.

Smith, N. 1988 'Lower East Side as Wild West: New City as New Frontier,' unpublished manuscript, Department of Geography, Rutgers University, New Brunswick.

Smith, N. K. 1971 *On Art and Architecture in the Modern World.* Victoria: University of British Columbia.

Squires, G. D. (ed.) 1989 *Unequal Partnerships: The Political Economy of Urban Redevelopment in Postwar America.* New Brunswick, N.J.: Rutgers University Press.

Strong, R. 1988 'Rus in Urbe,' Kevin Lynch Memorial Lecture, transcribed by P. Luck in *Urban Design Quarterly*, June 1989, 45–48.

Tafuri, M. 1980 'The Disenchanted Mountain' in G. Ciucci, F. Dal Co, M. Manieri-Elia, M. Tafuri, *The American City From the Civil War to the New Deal.* Cambridge, MA: MIT Press.

Tibbalds, F. 1989 'An overview,' *Urban Design Quarterly*, June, 1989, 40–44.

Trilling, J. 1985 'A future that looks like the past,' *Atlantic Monthly*, July, 28–34.

Walsh, A. H. 1978 *The Public's Business: The Politics and Practices of Government Corporations.* A Twentieth Century Fund Study, Cambridge, MA: MIT Press.

Wodiczko, K. 1986 'Public projections,' *October 38*, Fall 1986, 3–23.

Wolfe, T. 1981 *From Bauhaus to Our House*, New York: Farar, Straus, Giroux.

Wright, P. 1989 'Re-enchanting the nation: Prince Charles and architecture,' *Modern Painters*, vol. 2, no. 3, 26–33.

Zukin, S. 1989 'The City as a Landscape of Power: London and New York as Global Financial Capitals,' *Russell Sage Foundation Working Paper #3*, 112 East 64th Street, New York, NY 10021.

Zukin, S. 1991 *American Market/Place: Landscapes of Economic Power*. Berkeley and Los Angeles: University of California Press.

Seven

SOCIAL REPRODUCTION IN THE CITY: RESTRUCTURING IN TIME AND SPACE

Robin M. Law and Jennifer R. Wolch

University of Southern California

Restructuring is a term that promises to make sense of the profound reorganization in economic, political, cultural and social life which many have identified in the contemporary U.S. Yet most theoretical and research interest has crystallized around *economic* restructuring, or the way that production is organized. The fundamental changes in the organization of social reproduction (the reproduction of the labor force) have been relatively neglected, and the long-standing distinction between the two 'spheres' in analytic thought and empirical research has been replicated unchanged in much contemporary work.

We argue here that using social reproduction as a point of entry in an analysis of urban change is a valuable step toward developing an understanding of transformations in urban form and activity patterns in the U.S. It allows commodity production to be seen as part of a broad and shifting division of labor, which encompasses goods and services produced and labor performed in a variety of sites and conditions, including the home, factory, local voluntary organization, and state agency. It also gives weight to the changes (in labor pools, markets, and so on) that are driven by extra-economic structural forces such as demographic change. In that sense, we suggest, it is impossible to fully understand the restructuring of production without also understanding shifts in patterns of social reproduction. Finally, a focus on social reproduction is politically important, as it opens a political space for those whose primary *experience* of oppression is not felt through economic exploitation, by integrating their experiences into a larger and more inclusive analysis.

We begin our analysis by asking how the daily and generational reproduction of the (broadly defined) working class is achieved, and we identify a variety of resources involved in this process. These include state support, income from wage labor or informal sector trade, community services (including support from neighborhood social networks and charitable or voluntary organizations) and support from relationships with household members and kin. The ways in which people make a living by patching together support from these sources represent adaptive strategies in the context of social, spatial and temporal constraints and opportunities. These practices embody power relations of class, gender and race but are also sites of struggle and renegotiation that change over time, with different implications for different social groupings. In addition, they contribute to the transformation of urban structure, through the patterning of activities in time and space, expressed in the construction of places that serve as sites for these activities and that come to be imbued with meaning through their use.

We thus identify *four central institutions* as the basis for social reproduction, namely, the economy, the state, the community and the household. To understand how each is restructuring, we identify changes in how the institution is organized; how goods and services are produced and distributed; the conditions under which work is done; who has claims on support; and what relations of power derive from the organization of work and distribution of resources. The discussion of each is necessarily brief, but some of these issues are elaborated in further detail in the next section.

The second section of the chapter identifies *five trends in urban life* which are contributing to new activity patterns in time and space and which are, in turn, transfiguring urban space by creating multiple employment nodes and by shaping temporal rhythms, movement corridors, market demand and the meaning of particular spaces. Although by no means exhaustive, these five trends illustrate the way that the multiplicity of dynamics identified in the previous section interact to affect the daily lives of urban residents. Throughout the chapter, we focus on women and on low-wage workers as pivotal subjects in the new urban space: both as victims of restructuring and as active agents shaping space and time through adaptive strategies in their daily lives.

Restructuring the economy

Declining profit rates in the U.S. in the 1960s (as productivity growth failed to match wage growth) marked the beginning of a crisis in the economic system. By the early 1970s, it was becoming clear that the role of the U.S. in the global economy was changing, signalled by the collapse of the Bretton Woods agreement, which had preserved the dollar as the world's reserve currency. The response of transnational corporations to the low labor costs and inviting subsidies offered by developing countries desperate to encourage investment meant that an increasing percentage of the world's manufacturing and assembly operations was taking place in Third World countries whose role was once limited to supplying raw materials and markets. But this did not represent merely a switch in the relative position of *national* capital groupings: instead, more power was concentrated in *transnational* corporations, and the flow of capital across international money markets was facilitated both by this centralization and by the new communications technologies which helped establish an integrated global financial system.

At the same time, beginning in the second half of the 1960s, the relative social consensus underpinning the mode of regulation was challenged by resistance to the war in Vietnam, the student revolt, the claims of ethnic minorities and the women's movement. Some of this was driven by demographic forces, such as the impact of the large 'baby boom' cohort on the labor market and popular culture. In these social movements we see also a spurning of the mass-market mode of consumption and rejection of the ability of the homogenous American Dream to satisfy the diverse social groupings now making their voices heard. These challenges were met by a combination of responses, including the incorporation of marginalized groups into the more privileged segments of the labor market through programs such as affirmative action. In another response, the demand for more personally satisfying and non-alienating ways of living was translated into the marketing of commodities that formulated those inchoate stirrings as specific desires for a range of finely differentiated (and constantly outdated) commodities with an ever more carefully calculated symbolic value (see Chapter 1). But the oil price increases of 1973 and the ensuing stagflation revealed the limits of the adjustments possible under the old regime. Thus in the late 1970s the elements of what has been called a new 'regime of accumulation' (and its accompanying social and political 'mode of regulation')[1] were becoming clear (Aglietta, 1979; Alliez and Feher, 1987).

Among the chief features of the emerging systems were:

(a) massive waves of financial mergers and acquisitions, and the emergence of huge transnational conglomerates;

(b) rapid growth in sectors which incorporate flexible production techniques, including high-technology manufacturing, producer services and design-intensive craft-specialty manufacturing;

(c) the development of flexible technologies permitting short production runs and rapid turnaround times from design to final delivery;

(d) closer monitoring and more finely tuned methods of inventory control and ordering, including 'just-in-time' delivery systems and point-of-sale stock

[1]The French Regulation school (Aglietta, 1979; Lipietz, 1986) developed the concept of 'regime of accumulation' to distinguish periods and societies under capitalism in terms of the specific pattern in which the economic surplus is allocated between investment and consumption. The 'mode of regulation' consists of the institutions, norms and social practices which together help to sustain and stabilize a given regime. The regime of accumulation in the U.S. and some other developed capitalist countries since World War II is known as 'Fordism' and is characterized by mass production coupled with high mass consumption. The accompanying mode of regulation is built around a Keynesian welfare state. Fordism began to show signs of collapse and transformation in the early 1970s.

control;

(e) distinctive types of inter-firm relationships, including more extensive vertical disintegration with sub-contracting relationships, where transactions are conducted on the market rather than internally;

(f) flexible use of labor, including expanded use of part-time, temporary, and contract workers; and

(g) targeting of more precisely defined market segments rather than mass markets.

In this discussion, the focus will be on the new global and social division of labor which contributed to sectoral shift; the extension of subcontracting arrangements; and the growth in flexible employment practices and how these have altered the role of wage labor as a source of income for social reproduction.

In the U.S., those sectors which did experience rapid growth in employment included high-technology manufacturing (supported by extensive defense-related government investment); services which could not be relocated (e.g. retail, health care); and those producer service jobs associated with the role of the U.S. as the site of research and design, finance and management activities of transnational corporations. But the central feature of the changing industrial mix in the U.S. in the last decade has been the decline of manufacturing employment relative to services. While this has been a long-lasting trend, the symbolic turning point in many cities came during the 1980s when the number of people employed in the service sector (as defined by the Census Bureau) exceeded those in manufacturing.

The rise in service jobs can be attributed to shifts in the division of labor between nations; between firms; and between the household and the market. The new global role of the U.S. from the 1970s as the locus of financial and management activities in an internationalized production process, the 'New International Division of Labor' (Cohen, 1981; Frobel et al., 1980), was matched by an intensified polarization in wages and a profound spatial, political and cultural separation among workers. Within the U.S., vertical disintegration of formerly integrated large firms—part of the trend toward flexible specialization—meant that services once provided in-house were now increasingly purchased on the market (often from dependent contractors). Both of these trends contributed to the surge in the producer-services sector.

The growth in consumer services was generated as work done under wage relations of production (either for sale on the market or as part of the public sector), replacing work that was formerly done in the household (usually by women) without direct compensation. Another contribution to the growth in consumer services has been the trend toward privatization of state services, which has involved both traditional private-sector firms and increasingly powerful non-profit service providers. The commodification of household activities is driven by the coming together of two forces: the insistent search for new sources of accumulation inherent to capitalism, and the struggle by women for greater economic autonomy (resulting in both a pool of 'appropriately gendered' labor for these jobs and a demand by households for market provision of the services). Growth in mass-market consumer services (such as fast-food establishments) accounts for a large share of consumer service job growth. But there has also been significant growth in consumer services tailored to the demands of specialized market segments, usually provided at a higher price and marketed with the implicit promise of setting the consumer above the mass.

The shift to service-sector employment contributed to income polarization in the labor market: while many of the manufacturing jobs which were eliminated were in unionized, relatively well-paying industries, the new service-sector jobs tended to be in non-unionized industries, in sectors (such as health care) with relatively fewer jobs at mid-level incomes, and thus with a work force that was more sharply segmented around two poles in terms of working conditions, internal promotion prospects, social characteristics of workers and so on (Christopherson, 1989; Sassen, 1984a).

The extension of subcontracting through the vertical disintegration of firms has also contributed to an increased segmentation in the labor market. In essence, sub-contracting involves the downward shifting of risk; the constant costs of wages and overheads are borne by subcontractors in periods of reduced demand. Hence employment in subcontracting firms is likely to be less secure than in vertically integrated firms with the internal labor market characteristics of large corporations in the Fordist period. In some cases, especially for professional workers with skills much in demand, insecurity may be compensated for by higher wages. But the existence of many small firms, competing for contracts on the basis of cost in a market increasingly controlled by a limited number of clients, tends to drive wages down. In the absence of productivity gains through capital investment, firms resort to simple exploitation to wrest profits from the transaction, especially where the job requires low skill. In addition, the balkanization of the labor force into numerous small establishments makes unionization and concerted industrial action difficult. For example, the industrial structure of the garment industry in Los Angeles, which grew rapidly in the 1980s, increasingly comprises a large number of small sub-contracting firms who compete with each other for sewing contracts from the few powerful and financially secure manufacturers, who perform the design, cutting and marketing functions. The basis for competition is generally cost, which all too often translates into below-minimum wages for the (largely immigrant Latina) seamstresses in the small sweatshops (Scott, 1984).

The phenomenon of sub-contracting takes another form in the use of 'contingent' labor such as temporary, part-time and short-term contract jobs. The growth in temporary work has outstripped that of other forms of alternative work schedules since the 1970s. Employment in the temporary services industry has averaged an annual growth of 11 percent per year since the mid-1970s, compared to 2.1 percent for all non-agricultural jobs (Appelbaum, 1987). In addition, many firms (including all levels of government) are hiring temporary workers directly and preplanning such practices instead of relying on temporary workers only to fill unforseen demand.

Part-time work is also a way of dealing with fluctuating demand; when peak demand times are predictable, shifts can be organized to match them. New forms of part-time work are being developed that combine short working hours with variable shifts, such as the phenomenon of being 'oncall.' The rate of growth in part-time work from the 1950s to 1977 increased to twice the rate of all jobs, but, since 1977, the proportion of part-time jobs has remained fairly static (Appelbaum, 1987). However, what has happened is a shift both in the average hours worked by part-time workers and in the share who are working part-time involuntarily. The share of women working 15 to 29 hours a week has increased, while those working 30–24 hours has declined, a shift that is probably related to benefit packages. More dramatically, the percentage of part-time workers who would prefer to have a full-time job has grown; today, nearly one in twenty part-time workers would rather have a full-time job (*Wall Street Journal,* 1990).

Flexible employment practices allow firms to cut their labor costs by hiring workers only for those periods when the work load is high. But, perhaps more important, these practices also generate significant savings on non-wage benefits (pensions, health insurance, unemployment insurance fund contributions). One estimate by the Labor Department found that hiring contingent workers saved companies 15 to 24 cents on every payroll dollar in 1988 (Smith, 1990). Although in the past temporary and contract workers in jobs with full-time equivalents have tended to be paid a premium in their hourly wages to make up for the insecurity of employment and lack of benefits, this practice could well disappear in the current harsh economic climate. As new technologies are introduced that allow for the reorganization of job definitions and standardization of tasks which formerly required initiative and trust, more aspects of the work process (especially clerical) are likely to be transformed into temporary and part-time jobs. As Baran (1985) shows in her study of the insurance industry, a lowering in wage rates together with job redefinition is often associated with the opening up of jobs to women.

The outcome of these three trends (sectoral shift, sub-contracting and flexible employment practices) has been to segment and divide the labor market in new ways, which are as yet not clearly understood. Current formulations of labor market structure suggest that restructuring has produced a more complex and differentiated set of segments than before. The notion of a primary and secondary labor market (with the primary segment subdivided into independent primary and subordinate primary components), which generally coincided with divisions in wage, skill and security characteristics of jobs

(Doeringer and Piore, 1971; Gordon et al., 1975; 1982), can no longer be usefully applied. While a privileged group of permanent, full-time workers in secure jobs with good wages, benefits and internal labor markets can still be identified, they are now complemented by a variety of 'secondary' segments, including, for example, well-paid professional temporary workers; low-paid clerical workers once directly employed but now formally defined as independent sub-contractors; skilled but part-time workers in positions which are designed for high turnover (such as college students); and so on.

The correlation between labor market segmentation and (a) social characteristics such as gender and race and (b) industrial sectors is also being subverted. Although white men continue disproportionately to dominate the privileged 'primary sector' jobs, women are increasingly breaking into some of the professional and executive ranks. Yet new forms of segmentation emerge to express continued gender inequality. Christopherson (1989) notes that, within some skilled occupations, women work fewer hours (and thus earn less) than their more well-connected male counterparts, and women managers are more likely to be in small firms. And while secondary labor market conditions were once most notably concentrated in firms in peripheral sectors of the economy (highly competitive and labor intensive), this is changing. Increasingly now we see the emergence of 'hybrid firms' (Noyelle, 1987), where both primary and secondary labor market employment conditions are present among the direct employees, producing a more polarized job structure.

Given these developments in the labor market, workers are becoming less able to rely on their own jobs as secure and adequate sources of income to support themselves and their children. In this economic climate, the ability to rely on other sources (such as state support during unemployment or the income of a spouse) becomes ever more important. Yet restructuring in these institutions is also affecting the ability of individuals to draw upon these alternative resources.

Restructuring households

The diversity of household types and activity patterns within households has changed dramatically, but the effects express long-standing demographic and social trends. Longer lifespans (among whites) and declining fertility levels have meant that more people live into old age, where they are likely to live as couples or single households rather than in extended multi-generational families. Sexual liberation movements contributed to the rise in unmarried heterosexual and gay couples and to the reduced stigma of divorce. Increasing educational and employment opportunities for women and contraceptive and other medical advances facilitated delayed childbearing. The demographic bulge of the baby boom cohort has affected the mix of households as it matured, contributing first to a large share of families with children, later to an increase in the share of single households and, most recently, to an increase in child-bearing couples. The outcome of these trends is that there are now relatively fewer families which consist of the classic stereotype of employed father, housewife-mother and dependent children. In their place are a wider variety of diverse households.

Some figures bear out the scale of the changes (all are from Chen, 1982). One-person households almost doubled between 1970 and 1980, rising from 11 million to 18 million. In part, this is due to later marriage. The proportion of women between the ages of 20 and 24 who were single grew from 28 percent in 1960 to 50 percent in 1980, and a similar trend was observed in the ages 25 to 29. Also, divorce rates more than doubled in the same period, from 2.2 to 5.3 per 1000 population, while the remarriage rate declined. The number of single-parent households rose simultaneously. Twenty percent of all children were living with only one parent in 1980, compared to 12 percent in 1970. Most of these single-parent families were headed by a single woman.

Household structure was also affected by ethnic composition and the socio-economic position in which ethnic groups were inserted into the labor market and civic life. Immigration from Central America and Asia (linked in part to U.S. military intervention and investment, as Sassen (1988) points out) has contributed to a greater degree of ethnic diversity, especially in those cities such as New York and Los Angeles which have received the lion's share of immigration. Since a large share of these immigrants entered the U.S. either under the family-reunification policies, which privileged immigrants with immediate family members already in the country, or as undocumented entrants who drew on kinship linkages to survive in a hostile environment, immigrant families today are more likely than natives to form households with multiple members.

The most important shift in the activity patterns within households is due to the increased participation by married women (and especially, mothers of small children) in the paid labor force. One-worker husband-and-wife families declined from 43 percent of all households in 1960 to 25 percent in 1975, while two-worker husband-and-wife families increased from 23 percent to 30 percent (Chen, 1982). By 1980, 57 percent of the mothers of children under age eighteen were in the labor force (Chen, 1982), and by the end of the 1980s over half of the mothers of small children (under 6 years of age) were in the labor force.

Sectoral economic trends made it easier for women to move into paid employment. Many of the fastest-growing occupations and industries were those which were dominated by what had come to be defined as 'women's jobs': clerical, retail and personal service jobs, characterized by low wages, less than full-time employment, and little access to promotion. The sectoral shift identified above meant that these 'women's jobs' grew relatively faster than many other male-dominated jobs. From the point of view of the working class as a whole, this involved an overall loss in well-paying jobs, but the effect was experienced very differently by men and women. For women, the expansion in job opportunities in their traditional fields of competence—albeit poorly paid jobs—was favorable, making it slightly easier for them to compete with men in the labor market. And since this occurred at the same time as many women were gaining access in greater numbers to the occupations traditionally dominated by men, job opportunities expanded. The result (along with other factors, such as longer hours worked by women) was that the disparity between the average wages earned by men and women narrowed. Between 1973 and 1986, the position of women in the labor market improved moderately relative to men and also improved in absolute terms (Levy, 1988). But the narrowing of the wage gap should not be exaggerated. In 1973, women aged 35 to 44 in full-time, year-round jobs earned 54 percent of their male counterpart's annual earnings; by 1986, they still earned only 64 percent. The relative wage gains for younger women were slightly higher but were lower for older women (calculated from Levy, 1988).

The growth in women's labor force participation was accompanied by an increase in the number of households (especially families with children) headed by single women. Households headed by a single woman under age 65 grew by 67 percent between 1973 and 1987, while those with conventional husband-wife couples grew by only 7.1 percent (Levy, 1988).[2] From one perspective, the formation of these households could be seen as a response to the improved opportunities for economic self-support. More women in unbearable marriages could consider divorce knowing that they could feasibly support themselves and their children through employment. Obviously not all divorces were initiated by women seeking independence, but there is some evidence of the link between the divorce rate and the extent of economic opportunities for women (Hartmann, 1987). For whites, divorce and separation were the main reasons behind the formation of single-parent families; by the 1980s, 85 percent of all white single mothers had once been married (Mulroy, 1988).

Among blacks, a larger share of single mothers had never been married; the share of never-married black women between 18 and 44 years grew from 27 percent in 1970 to 44 percent in 1982 (Mulroy, 1988). This can be interpreted in the light of the economic support that men could offer as husbands. Younger men without college education were the group whose wages showed the least growth since 1973 (Levy, 1988), and young black men have borne the brunt of the shifts in occupational and spatial labor demand. Their marginal position in the labor market translates into a marginal role in the household structure, which, some have suggested, has lowered marriage rates (Wilson,

[2] However, some of the explosive growth in female-headed families is the artifact of a change in the way that the Census Bureau measured female-headed families living in a household with other relatives.

1987; Holloway, 1990). In short, for both blacks and whites, one effect of the feminization of the labor market has been to fracture the economic basis of marriage.

Along with the feminization of the labor market came another pattern: the feminization of poverty (Sidel, 1986). The persistence of the gender gap in wages; the inequitable division of assets at the time of divorce; and the ineffective implementation of child-support agreements mean that female-headed households are one of the fastest growing groups among the poor.[3] Female-headed families grew from 11 percent of all families in 1970 to 14 percent in 1978. Single mothers had a much higher labor-force-participation rate (67 percent in 1980) than mothers in two-parent families (54 percent) (Chen, 1982). But the median income for female full-time, year-round workers heading families was only 59 percent of what was earned by similar male-headed families. Thus they increased from 21 percent of the poor in 1960 to 44 percent in 1981 (Kuhn and Bluestone, 1987). The result is that more families containing children are burdened both by low income and the heavy unremitting demands for time and energy which fall upon the single parent, and more children are growing up in poverty.

These shifts in household composition and income suggest ways in which the function of the household as a redistributive mechanism may be changing. It appears that women and children are now less able to make effective claims on the wage of a husband/father for support, due to both the reduced relative earning capacity of men and the increase in divorced men who pay no alimony or child support. While married men now do slightly more domestic work, it is clear that on the whole the household still functions as a means by which men make claims on women for domestic support through direct labor. Reciprocal claims between parents and adult children are also changing. Longer lifespans have meant that many adults in the generation where both men and women work fulltime are facing conflicts between the time demands of their job and the responsibility for caring for aging parents. And as income growth stagnates, the expectation that children's financial success will exceed that of their parents diminishes. Instead, existing class patterns are entrenched and widened as those adult children with parents who can afford to, for example, contribute to home purchase are privileged above their generational peers.

[3]Mulroy (1988) summarizes data on the economic impact of divorce. In most marriages there is minimal accumulation of assets; only 36% of divorced women were awarded a property settlement in 1986. The most significant asset is the husband's earning capacity to which the wife contributed, and in 1985, only 15% of the ever-divorced or currently separated women in the country had an agreement or award to receive alimony. So it is not surprising that women's financial position worsens after divorce, and this is exacerbated if there are children of whom she receives custody. In 1985, only half of all the women due child-support payments received the full amount due (the average amount of the award was $2,000 per year), and one quarter received nothing. A study in California of the late 1970s (Weitzman, 1985, cited in Mulroy) showed that in Los Angeles county one year after divorce, the standard of living of mothers and children had declined by 73%, while that for fathers had risen by 42%.

Restructuring the welfare state

The function of the state as a source of social services has been transformed, predicated on a neo-conservative economic doctrine which prescribed limits to wage increases, rising interest rates, tax cuts and a decline in government spending. In essence this has involved an abandonment of many of the inclusive, redistributive aspects of ruling ideology and the promotion of a more brutal individualism allied to libertarian themes. The transformation in the role of the welfare state will be described with reference to the following themes: the elimination of programs and cutbacks in eligibility; the regressive nature of those redistributive programs that were retained; the shifting of federal responsibility to state and local governments; the privatization of services through contracting out and through extension of the voluntary sector; and the emergence of new social needs that were unmet. Each of these trends served to fragment and divide individuals, to heighten differences rather than level them.

Government spending on social welfare in the U.S. remained fairly constant at about 10 percent of GNP from 1940 to 1960, but during the 1960s and 1970s it expanded dramatically as the War on Poverty was launched. By 1975, public expenditure accounted for 20 percent of GNP (Gilbert, 1983). Between 1967 and 1978, real cash transfers per household increased by 67 percent, while real GNP grew by 9 percent (Gottschalk and Danziger, 1984, cited in Danziger and Feaster, 1985). The cutbacks in public spending began in the late 1970s and continued through the 1980s. By 1979, the share of GNP made up by public spending on social welfare was 18.5 percent (Gilbert, 1983). Between 1978 and 1982, real cash transfers declined by 1 percent, while real GNP declined by 7 percent (Gottschalk and Danziger, 1984, cited in Danziger and Feaster, 1985).

The rise in social welfare expenditure had been particularly concentrated at the federal level and had involved a broadening of the eligibility requirements toward more universal entitlement programs that provided benefits to the middle class as well as poor. From 1973 to 1976, the share of the total federal social welfare budget that was not solely directed at the poor increased from 46 percent to 54 percent, even while official poverty rates were climbing (Gilbert, 1983). When the cuts in spending came, they did not affect the universal entitlement programs such as Medicare, where spending kept rising, but were aimed at discretionary programs such as housing programs. The attack on income support for the poor was so prominent that this period has come to be called the War on Welfare. The median value of AFDC (Aid to Families with Dependent Children) benefits and the level of General Assistance (welfare relief funded by counties) declined by about one third in constant dollars between 1970 and 1985 (Katz, 1989). At the same time, subsidies to the middle class such as income tax relief on home-ownership mortgage payments were retained.

At the same time that state services were being retracted, new populations in need were emerging for which minimal state support was available. Among these we may point to displaced homemakers; substance abusers; and people with AIDS. Although the elderly as a whole were relatively well protected by Social Security from the cuts in human services, increasing longevity has meant a rise in the number of very old who need care by others. And at the other end of the age spectrum, the lack of any state system for child care for children under school age has emerged as a striking problem in family life as more mothers are forced into the labor market to support their families.

The cuts in federal spending were accompanied by the 'New Federalism,' which shifted responsibility for many programs to state and local governments, many of which were without the income resources to make up the shortfall. So, as the federal government cut social programs, many states introduced more restrictive rules for eligibility to welfare rolls and city governments closed down services. The transfer of federal responsibility to state and local governments accompanied and stimulated a shift towards privatization. State and local governments charged with new responsibilities turned to the voluntary sector and for-profit providers (Wolch, 1990). Privatization had been proceeding at different rates in various parts of the welfare state, and voluntary organizations were becoming more dependent on purchase-of-service agreements with government. By 1980, private contracts made up about one-third of the total revenues of voluntary social service organizations (Gilbert, 1983).

Although the cuts in social welfare spending hurt many voluntary organizations, they also stimulated a reorganization and a move into new forms of revenue raising and service provision. The result was that the spending cuts sharpened the role of the voluntary sector as a shadow state (Wolch, 1990, p. 74) and generated a more complex and contested relationship between state and voluntary sector. Cuts in federal spending (along with local cuts such as California's Proposition 13 tax limitation) also resulted in job losses. The groups hardest hit were women (who are concentrated in social service professions) and minorities, especially African-Americans, who have found public administration to be one of the sectors where the effects of racial discrimination on income are lowest (Bluestone et al., 1973).

The elimination of programs and tightening of eligibility, the shifting of responsibility from federal to lower levels of government and the privatization and 'voluntarization' of services each contributed to a redefinition of citizenship in terms of claims on the state for help with social reproduction. The combination of all served to further undermine the myth of inclusion and of a homogenous political subject (with varying needs but equal claims) which ruled during the Fordist period. Cutbacks in spending through tightening eligibility requirements (for example, through a more restrictive definition of disability during the Reagan years) served to shrink the pool of 'deserving poor' by redefinition. An ideology of sturdy independence was promoted to diminish claims from the class including the most needy at the same time that other state policies (such as tax reforms) redirected income to those least in need.

The shift to local state spending also served to undermine the definition of a *national* political subject in favor of the local and regional. This limited the redistribution of income between rich and poor areas of the nation and also tended to undermine the position of blacks, who have historically suffered from greater institutionalized discrimination in services from local than from the federal government. In addition, the argument of limited resources could be more truthfully claimed in the context of state and local budgets than in the federal government, where spending in sectors such as defense was ballooning.

Finally, the privatization of services also altered the relationship of service providers to recipients. Clients, once defined as political subjects (e.g. as potential voters, or citizens entitled to non-discriminatory treatment) became redefined as consumers. Alternatively, their access to services was mediated through their membership in the community supplying the service or their status as potential members. People seeking help in social reproduction were thus faced with a new set of requirements (such as proof of state residency or willingness to join in communal prayers) before they were granted access, and this generated a new geographic and social distribution of resources. For example, the needs of people afflicted with AIDS have been met largely by voluntary organizations created in the gay white male community, resulting in deprivation by those groups outside that community without comparable resources.

The outcome is not only that people are defined in terms of their membership in social groupings, but also that people in need of support come to define themselves and draw on their 'personal capital' to position themselves and act in ways to obtain these services. Thus welfare state restructuring is one more way in which social groups become fragmented and in which people draw on their personal resources to develop a 'flexible response' to a changing environment of need and structured access to resources for reproduction.

Restructuring of communities

Residential communities remain highly segregated by race and class, while stages in family life-cycle and immigrant origin are still strong factors in sorting out spatially defined communities. But spatially defined urban residential communities are affected by two trends: first, a more-self-conscious, clearly defined segmenting of spatial communities; and, second, a greater use of lifestyle and what may be called consumer identity as the basis for the formation of a community. In both of these, capital plays a major role.

The concentration of financial resources into fewer firms has facilitated huge multiple-use developments, with highly sophisticated marketing strategies targeted at specialized sub-groups such as wealthy, active retirees. Supporting these investment strategies is a burgeoning industry of computerized analysis of demographic and consumption patterns which has already resulted in detailed mapping by postal code zone of the entire U.S. (Weiss, 1988; see also Chapter 8). In a sense, then, households are defined by what they buy; specialized spaces are then constructed to serve these consumer groupings, and households are sorted into the spaces either by attraction (through marketing) or by exclusion (through, for example, redlining and the withdrawal of retail outlets from communities without the appropriate buying power). Even the most subversive social trends such as gay liberation can be, and are, translated into consumer demands which are then met through a marketing strategy.

Yet the impact of trends in capital accumulation can be contradictory; financial concentration has also stimulated the creation of a global consumer market that actively rejects *local* identity and loyalty in favor of a specialized niche in an *international* hierarchy of style and tastes. So while economic restructuring is one of the new forces by which spatially defined communities are being generated and maintained, the same process also serves to fragment existing communities and to break down the significance of local residence, and it operates in local areas in conjunction with particular local political and social trends. Thus an attempt to identify patterns in community restructuring yields evidence of trends toward both fragmentation and coalescence, sometimes simultaneously. Each of these will be discussed in turn below.

Fragmentation of local communities occurs as the fine web of long-established social networks among residents and people working locally is eroded. For example, financial concentration has meant that many local retail and other businesses are taken over by larger, outside corporations, sometimes to be closed down if the profit margin is less than can be gained elsewhere. But even if they remain in place, distant management is usually less responsive to local concerns. At the political level, cutbacks in local state services and marginalization of residents from local politics (as politicians become increasingly dependent on financial contributions from developers and other business interests) both eat away at the network of reciprocity which formed the basis of old-time patronage politics. Finally, trends among households also subvert the development of local community. Households respond to the mobility of capital by moving to find employment, and the mobility of the population fragments existing neighborhood support structures. As more women work outside the home, fewer are available to perform voluntary work in local community institutions as well as the less well-recognized work involved in building local friendships which cement neighborhoods. And in gentrifying neighborhoods, the outside 'consumer' orientation of new residents undercuts local businesses and institutions. Williams (1988) compares the community-building daily activity patterns of long-time residents in one Washington, D.C., suburb—epitomized in the 'street work' of adult men—with those of the new gentrifiers who send their children to private schools, who avoid shopping in local stores, and whose social networks are non-local.

But there is another set of forces that serves to coalesce neighborhoods. Economic restructuring has created more niches for small businesses, often in labor-intensive fields, in a dependent sub-subcontracting relationship to larger fractions of capital and operating on the fringes of legality. Although seemingly marginal, they are in fact central to the functioning of the emerging economic system. These niches are often utilized by ethnic communities or other groups with poor competitive standing in the formal labor market (such as unskilled mothers of young children). Recent immigrants draw on the resources they have, such as the ability to mobilize family labor and familiarity with a specialized market, to carve out an economic place. Since this often involves reliance both on labor drawn from the immigrant community and on the same community as a market, this economic place also becomes a space in the city associated with the immigrant residential neighborhood.

Women are also a significant fraction of those working in small businesses. The increasing need for two incomes to support most families has led to more women seeking income opportunities, usually close to home. Women's mobilization of local friendship networks to find jobs, to provide a market for a business operated from home (such as child care) or to provide the services that are needed to juggle home and work responsibilities may, in fact, strengthen local social networks. In this sense, then, the interaction between the coping strategies of individual households and the economic accumulation imperatives of global capital may generate a revival of local social networks, albeit built upon the most desperate struggle for economic survival by households.

The devolution of responsibility for state services from federal to local levels and from the state to the voluntary sector may also contribute to community formation by building up the importance of community-based service providers. The story of the provision of services to people with AIDS in San Francisco and Los Angeles (Shilts, 1987) is a classic example of a community-based effort—in this case, rooted in the white male gay communities, which were in turn rooted in particular spaces, such as the Castro district in San Francisco and West Hollywood in Los Angeles. Other experessions of locally-based action are the slow-growth and 'not in my backyard' movements of homeowners to resist commercial development and the human service facilities. In California these have emerged in recent decades as a significant form of political activism, albeit one with a long history in the region (Davis, 1990).

Finally, the networks of kin and friendship linking households that share ethnicity and lifestyle continue to generate spatially defined communities, although these may take seemingly exotic new forms with the arrival of new immigrant populations or the formation of new lifestyle-based social groupings. In many cities, the 'zones of transition' surrounding downtown continue to function as receiving areas for recent immigrants. But where the mass of the population are low-income renters, these areas are highly vulnerable to displacement through gentrification or commercial redevelopment. The communities that have the strongest and most permanent spatial identities are those with access to resources for capital investment in homeownership as well as in businesses. Yet this introduces cross-cutting class divisions, which can serve either to support or undermine their spatial integrity.

The situation of gay-identified communities is illustrative. In both Los Angeles (West Hollywood) and San Francisco, the formation of gay-identified spaces has been facilitated by both the capital investments of members of the community in businesses and homes as well as the economic stimulus of the high levels of disposable income that accompany households without the lower income-earning power of women or the financial costs of dependent children.[4] But not all members of the community are equally affluent: in West Hollywood, for example, almost 90 percent of residents are renters and over one quarter are elderly. The fragile alliance among classes is liable to be undercut by disputes rooted in divergent class interests which appear in other neighborhoods. For example, in West Hollywood, in one recent episode, local residents and business owners objected to the distribution of food to homeless people in a neighborhood park. Similar conflicts can no doubt be traced in immigrant communities, especially where different waves of immigrants have different capital resources (such as among the Cubans in Miami or the Chinese in New York).

The situation of African-American communities today shows how class division—in concert with continued racial discrimination—can fragment the spatial integrity of neighborhoods. Historically, the high levels of racial segregation (enforced through restrictive covenants and other legal mechanisms) meant that black urban communities contained a mix of all classes. However, low access to sources of capital limited community investment in the local neighborhood; those who attained middle-class status did so primarily through employment in public administration and professional services. In recent decades, three factors have eroded the social and spatial basis of the community. First, a trend of black out-migration from metropolitan areas (Johnson and Roseman, 1990) has resulted in the loss of homeowners and long-time residents. Second, a certain easing of the legally enforced housing segregation practices has allowed some middle-class blacks to move out of the ghetto, albeit to segregated suburbs (Kain, 1987). And third, the deindustrialization of inner-city industrial areas and systematic withdrawal of

[4]These spaces are largely identified as communities of gay men rather than of lesbian women.

investment from local properties has diminished local employment opportunities. Those who remain are thus likely to include a high proportion of people living in intense poverty and unemployment, a situation which has led to the wholesale labelling of the entire community as an 'underclass' (for the debate on the underclass, see Wilson, 1987; Hughes, 1989).

Urban activity patterns in time and space

In this section, five trends are described as examples of the complex outcomes of the restructuring processes outlined in the previous section. Each encapsulates some of the main issues in the reshaping of social reproduction. The focus is on the way that these trends affect the daily activity patterns of different groups defined by class and gender, and how these, in turn, generate new uses of space and forms of the urban built environment. We shall show how the emerging patterns of social reproduction represent adaptive strategies by individuals and households–flexible responses–to a changing context and mix of available resources. On occasion the conflicting interests and uneven distribution of costs and benefits is expressed in 'turf battles' over the use of urban space. Through a discussion of these five pivotal trends, some of the central features of the spatial organization and character of the post-Fordist city are brought to light. The case of Los Angeles is used to illustrate the argument, but many of the processes described for Los Angeles are replicated in other U.S. cities.

The five trends, which will be discussed in turn below, are: (1) polarization of earnings and household incomes; (2) marginalization of certain adults from both state support and labor market; (3) incorporation of more women into wage labor; (4) reshaping of domestic labor; and (5) increased flexibility and variety in the hours and conditions of wage labor. Associated with each of these are specific kinds of places and spatial processes and temporal rhythms of activity. The polarization of earnings is linked to the emergence of suburban employment nodes and gentrifying inner-city areas, with implications for commuting times and flows. The marginalization of a population segment is linked to the claiming of public space as a site of daily life for homeless people. More women in the labor force has meant a daytime depopulation (or, rather, defeminization) of the suburbs, with parallel growth in communities with many single mothers. The reshaping of domestic labor is revealed in the upsurge in sites of extra-household consumer services, contributing to new spatial patterns and temporal patterns of work and home life, more varied patterns of commuting and a blurring of the boundaries between home and work as both new and old technologies (from computer terminals to sewing machines) are introduced into traditional sites of reproduction.

Polarization of incomes

The distribution of incomes in the U.S. is becoming more polarized, contributing to a growth in the share of people living in poverty. While changes in state policies (such as the tax system) have contributed to this outcome, the most significant contribution is the polarization in the earnings of workers. A long-standing trend toward equality in earnings among full-time, year-round workers was reversed in the early 1970s. Harrison and Bluestone (1988) demonstrated that those full-time, year-round workers earning less than half of the median wage declined from 21 percent to 12 percent between 1960 and the early 1970s, as part of a steady trend toward equalization in salaries. However, in the early 1970s this pattern reversed, and their share of the labor force had risen to 17.2 percent by 1986. At the same time, growth rates in the high-wage segment also exceeded those in the middle-wage segment. Average earned incomes of full-time year-round workers have declined, and if part-time workers are added, the drop is more precipitous.

This polarization is translated into growing inequality in household and family incomes. The impact has been muted to some extent by the participation of more family members in the labor force, but a number of studies (Harrison and Bluestone, 1988; Levy, 1988) show that there has been a sharp trend toward greater inequality in family income since the late 1960s. At the lower income levels, this has been expressed as an increase in poverty rates. The proportion of the population living below the federally defined poverty line declined from 22 percent in 1960 to 13 percent in 1970 but then grew again to 15 percent in 1982 in the depth of the recession. By 1986 it had dropped only slightly, to 14 percent, and the absolute numbers of people living in poverty continued to grow (Katz, 1989; Danziger and Feaster, 1985). At the same time, median family income remained stagnant: after 26 years during which it increased by around 3 percent per annum, between 1973 and 1987 it grew by only a few dollars in real terms (Mattera, 1990). In part, the growth in low-wage jobs can be attributed to the impact of the baby boom cohort on the labor market, which swelled the number of young workers without experience and drove down their starting wages. However, several studies have shown that demographic explanations alone are inadequate (Sawhill, 1988; Harrison and Bluestone, 1988); a significant element has been the loss of well-paying manufacturing jobs in unionized industries and the growth of jobs in the service sector, where lower-skilled jobs have historically (in the U.S.) paid less than in the goods-producing sectors of the economy. Downward pressure on wages for low-skill jobs has been exacerbated by state policies (such as retaining the same minimum wage rate for eight years).

But the replacement of manufacturing jobs by service and retail jobs did not only impoverish lower-skilled workers in terms of earnings: it also impoverished them by removing many of the local, easily accessible job opportunities that had existed in the heavy industrial areas close to traditional working-class neighborhoods. The historical development of the Fordist city had been predicated on a spatial arrangement of proximity between blue-collar residential areas and the industrial zones where many of the residents worked. This symbiotic relationship was undermined by the long process of suburbanization of retail, industry and later of offices; the growing economic importance of retail and consumer services dependent on an affluent local market; and the deindustrialization which peaked in many cities in the recession of the early 1980s. Long-established working-class communities lost many of their local factory jobs, and the low incomes of the residents could not attract the retail and service jobs that were springing up in the outer suburbs. Inner-city communities became steadily more impoverished. Although the overall numbers of people living in poverty grew by 8 percent between 1970 and 1980, the people living in concentrated poverty areas (defined as places where over 40 percent were in poverty) grew by 36 percent (Sawhill, 1988).

The experience of Los Angeles reflects many parallels with other cities. Along with the emergence of nodes of concentrated poverty, nodes of concentrated wealth have also developed, epitomized by the gated communities with surveillance services and other private facilities. At a slightly larger spatial scale the most striking feature of suburban higher-income communities was the burgeoning of office parks, shopping malls and entertainment/recreation centers in their midst, which generated employment both for local residents and for low-skill workers from other areas. High-technology manufacturing industries have also tended to locate on the fringes of the metropolitan area, giving rise to new industrial nodes in places such as Chatsworth/Canoga Park (Law, Wolch and Takahashi, 1991). This more dispersed and decentralized distribution of jobs (Gordon et al., 1986) has been lauded as a resourceful market response to the congestion and commuting problems of growing metropolitan size, a natural 'evening out' of the jobs-housing balance. But while the spatial distribution of jobs may have become more decentralized, the spatial distribution of people (in terms of the location of their homes) has continued to be finely sorted by class and race. A decline in affordable housing, partly brought about by cuts in federal housing subsidies, tax changes, demolitions due to seismic safety standards and growth control, has meant that low-wage workers have continued to crowd into the inner city (Law and Wolch, 1991).

Yet the demand for affordable centrally-located housing is under pressure in Los Angeles, as in other cities, by competition from gentrification of local neighborhoods with the appropriately desirable housing stock, apartment construction and loft conversions in central high-density business areas (Giloth and Betancur, 1988; Smith and Williams, 1986). The development of these spaces is also a function of new household forms and new patterns of consumption within them. As Ehrenreich (1983) argues, the nuclear family suburban household around which post-war Fordist consumption was organized represented a limited local market for goods, a market which could be expanded with the establishment of more single-person households. In this vein, the emergence of the 'sophisticated bachelor' consumer ethic in the 1960s (epitomized by the lifestyle promoted by *Playboy* magazine) is central to the construction of an alternative consumption ideal. In the 1980s the two-career 'yuppie' couple played a similar role. For these households, higher income was not necessarily expressed in a demand for more space (as assumed in neo-classical urban economics) but could instead be expressed in demand for housing in close proximity to work and to densely developed sites of consumption.

As Sassen shows in her studies of New York (1984a, 1984b) the polarization of the income structure associated with the producer-services sector was allied to a polarization of consumption structure: the affluent households generated a demand for a range of consumer services involving low-wage labor or even informal arrangements, including restaurant meals, domestic cleaning, building renovation and personalized production of fashion clothing and jewelry. At the same time, the residential choices of the affluent households (e.g., for loft living) displaced traditional industries from downtown regions, as Giloth and Betancur (1988) show for Chicago, and displaced renters from low-income neighborhoods ripe for gentrification.

The new suburban industrial sites are also marked by a sharp disjuncture between the housing choices open to segments of their labor force, since the skill and wage structure of industries such as silicon chip manufacture is notably polarized. While the rise of these high-technology industrial complexes often stimulates or accompanies the establishment of local communities of low-skill labor (such as Vietnamese immigrants in Orange County), the lack of affordable housing coupled with the low wages paid to production workers generates substantial problems (Wright and Riave, 1989). So the polarization of incomes has been translated into a polarization of experiences: for those at the upper stratum, the opportunity to reduce commuting time by living close to work in an environment richly supplied with urban resources has been enhanced; while for those at the lower levels, local employment opportunities have declined, and the new jobs are located far from affordable housing. In the neo-classical urban economic model of spatial structure, which fairly faithfully reflected Fordist urban form, higher-income earners lived further from the densest node of employment (the city center) and travelled further to work each day. But the shifting location of housing and employment by class is transforming that simple model. In this process, a closer integration of home and work in urban space may be achieved, but one for which the costs and benefits are heavily skewed by class.

Marginalized adults

Adults of working age who had been eligible for some direct support by the state (such as unemployment insurance) were particularly hard hit by the contraction of spending on social services in recent years. While Social Security (which was index linked) protected many of the elderly from poverty, the holding down of benefits, tightening of eligibility, and total elimination of other programs involving transfer payments contributed to a rise in poverty rates among adults and the children who depended upon them. Among the victims were adults with disabilities. In the early 1980s, the Reagan administration tightened eligibility and removed about 200,000 people with disabilities (many psychiatric) from the Social Security insurance rolls (Katz, 1989). At the same time, the closing of institutions for the mentally disabled and the failure to provide adequate funding for the community care that was meant to replace them swelled the population of disabled adults who were unable to support themselves through wage labor but who were also often denied access to state support through barriers such as excessive paperwork requirements (Dear and Wolch, 1987).

Adults who were in the labor force but suffering from temporary unemployment also lost some of their access to direct support as unemployment insurance benefits were cut (Brooks and Ness, 1990). Between 1970 and 1985, average General Assistance (the relief provided by state governments) declined by 32 percent (Katz, 1989). Employment and training programs such as CETA and Job Corps that provided some work relief and assistance to structurally disadvantaged individuals in entering the labor market were also cut; federal obligations for employment and training programs dropped by about 65 percent between 1981 and 1983 (Palmer and Sawhill, 1982) as part of the Reagan budget cuts. Aid to Families with Dependent Children benefits were held down too between 1970 and 1985, resulting in growing poverty among recipients. By the early 1980s, 'new-style workfare' programs had been introduced in many states, and one of the requirements was that mothers of children between the ages of 3 and 6—once accepted as fully engaged with child care—would be required to work (Katz, 1989).

Thus an ever-larger number of people on the margins of the economic system who were once entitled to some level of state support (including assistance in entering the labor market through state-run education and hiring programs) have now been redefined as potential workers who must bear the burden of full or partial support through their individual efforts. One result of the drop in state support has been to swell the number of people competing for low-skill jobs, which has tended to hold down wages and thus raise the numbers of the working poor. But at the same time, the changing economy has resulted in declining opportunity for many of the same groups, resulting in simultaneous marginalization by state and labor market.

Sectoral and spatial shifts in employment, expressed in trends such as suburbanization, deindustrialization and cuts in state employment (which had historically provided good jobs for minorities) generated concentrated unemployment in the inner city. Younger workers with less than a high-school education were the ones who suffered the most from the slowdown in wage growth in recent decades (Levy, 1988), and minorities were particularly hard hit. As the economy changed, the number of jobs requiring physical strength and repetitive manual labor shrank, to be replaced largely by sales and service jobs requiring social skills and personal qualities defined as attractive by the dominant culture. This expanded opportunities for groups such as middle-class women but penalized groups such as young black men. Deprived of some of the state support through work relief such as the Job Corps, when funding for those programs was cut, they were also now largely deprived of the opportunity for earned income as entry-level jobs in warehousing, transportation and manufacturing moved offshore or into the suburbs.[5] Whereas, in 1955, 66 percent of the 18- to 19-year-old black men had been in the labor force, by 1984 only 34 percent were, yet at the same time the rate for white men dropped only slightly, from 64 percent to 60 percent (Katz, 1989). By the late 1970s, the percentage of black males with *no* wage and salary income had grown to 11 percent, compared to 6 percent in 1968 and 4 percent for whites in 1978 (Saks, 1983). These marginalized workers contribute to the pool of 'discouraged workers' (about 10 to 20 million people) who have stopped looking for work after long unemployment. The group has remained fairly stable in size since the early 1980s despite apparent labor shortages and declining unemployment rates (Christopherson, 1989).

How did the growth in these marginalized populations affect patterns of social reproduction in U.S. cities, and what did this mean in terms of urban space? We suggest that the dramatic rise in the homeless population and the conflicts over use of public space which have accompanied the phenomenon are the most significant expression of this trend.

First, in the absence of adequate support through employment or state services, informal activities (including predatory crime) have grown in the poorest areas. Shopping areas, parks and street corners have become the sites of informal trade (ranging from food to bootleg cassettes to prostitution and drug sales). Recycling centers that purchase aluminum cans and medical firms that pay for blood are now common elements in the local survival economy. In the backyards of modest homes, garages and outbuildings are quietly converted into makeshift apartments. A study in Los Angeles suggested that in

[5]The extent to which the unemployment of black youth is caused by spatial mismatch between jobs and labor is subject to debate, with some studies showing that it is 'race, not space' which has been crucial (Ellwood, 1986). But the general point about the disparity in employment opportunity between the (mainly minority) inner city and (mainly white) suburbs holds (Kasarda, 1989).

1987, as many as 200,000 people were living in this 'garage housing' (Chavez and Quinn, 1987). Illegal subtenancy and doubling up of families also grew: Hopper (1990) reported one study of New York public housing that estimated that half of all units had additional tenants.

Second, greater demands are placed on employed friends and relatives, even as these individuals themselves suffer from slow wage growth and increasing poverty. As Katz (1989) points out, this dependence on a network of kin and others is one of the most important economic strategies of the poor, yet very little is known about it. Exchanges of information, favors and the payment for services form part of a web of social relationships that hold together and sustain low-income communities (Williams, 1988). Help from friends and relatives is crucial to the survival of the very poor; a recent study by Husick and Wolch (1990) shows that the absence of kin is one of the markers that distinguishes homeless applicants from other applicants for indigent relief. But for those on the margins with little to exchange, such resources are often stretched to the limit. So, for example, survival strategies among homeless people involve a pattern of intermittent and varied use of a range of resources and self-provisioning. As Rowe and Wolch (1990) show in their study of homeless women, the maintenance of social networks is a crucial aspect of their social reproduction and structures their daily activity patterns and use of urban space.

But while the informal sector activities in the homes and residential neighborhoods of the poor are largely hidden from the eyes of the rest of the city, a third survival strategy of marginalized people is highly visible and has contributed to the physical reshaping and political redefinition of certain urban places. It is this most desperate manifestation of 'informal housing provision'–people living out of shopping carts, in cardboard boxes–which has provoked the most intense conflicts over the use of space and which has marked the post-Fordist city most unmistakably, creating what Dear and Wolch (1987) have called 'landscapes of despair.' Homeless people have quietly made spaces for themselves all over the city: in airport terminals, freeway underpasses, suburban branch libraries, neighborhood parks and highway rest areas. But their presence downtown–in the heart of the spaces claimed by finance capital–has been the most visible and has been the impetus for some of the sharpest conflict over the use of urban space (Roth, 1990 on Seattle; Davis, 1990, on Los Angeles; Gleeson and Wolch, 1989, on Venice, California).

In Los Angeles, the simultaneous rise in the homeless population and the burgeoning of the central business district are not unrelated; the global shifts in the economy that threw some people out of work and onto Skid Row also generated new jobs for executives in finance, international trade and legal services in the skyscrapers of the revitalized downtown. In a world of spatial disjuncture where a decision in one boardroom is felt in a factory closure many miles away, this spatial proximity of victims and beneficiaries of the new economy is unusual. The presence of homeless people in the shadow of the skyscrapers presents an obstacle to the presentation of the local boosters' image of downtown and threatens the property values and profit rates of local retailers and manufacturers. The conflict is fought out through indirect measures that have reshaped the urban environment: removal of public toilets, redesign of bus benches and heightened surveillance (Davis, 1990) as well as direct intervention, such as intermittent sprinklers set off during the night in local parks, bulldozing of settlements or selective enforcement of jaywalking legislation by police.

But the flood of homeless people to the missions and social services in Skid Row not only conflicts with the image of the city as a space of international capital investment, it also interferes with attempts to develop downtown as a site of social reproduction for the high-income financial sector work force. The insistent daily presence of homeless people and their use of the public spaces of downtown (to sit, sleep, panhandle and search through trash) has challenged the possibility that those spaces may become sites of social reproduction (theaters, open-air restaurants, etc.) of the desired high-income residents. Thus while the conflict between homeless residents and local property owners can been seen as a confrontation between use value and exchange value of downtown space, much of the potential exchange value of the space is predicated on its role in the grand strategy of developing downtown to meet the social reproduction needs of a particular affluent urban population. In this urban microcosm, we see how the conflicting social reproduction practices of different classes are translated into conflicts over the use of urban space.

Incorporation of women into the labor force

The proportion of women engaged in paid labor was around one-third of all adult women during the 1950s, but it rose dramatically during the 1960s and 1970s, especially among women in the 25- to 44-year-old age group. Whereas the labor force participation rate for this group was 48 percent in 1970, by 1985 it had risen to 71 percent, and the equivalent rate for mothers of small children had risen from 40 percent to 53 percent (Christopherson, 1989).

There are three ways in which urban space has been altered by this trend. First, the demise of the full-time housewife has transformed the nature of single-family surburbia. Second, women's employment may have affected the locational decisions of male-female dual-worker households and overall journey-to-work patterns, given the distinctive spatial and temporal features of women's employment. And third, the rise in households headed by employed women has contributed to the emergence of distinctive urban zones.

Wage labor by women undermines the distinctive role of the suburban community around which the Fordist city was constructed. This is not to say that the low-density, single-family-centered, racially and income-segregated residential communities do not survive. But without the presence of full-time housewives, their functioning changes. Women are less available to perform the unpaid volunteer community services (Daniels, 1986) and build the spatially bound friendships that create and maintain a local community.

In a sense, then, the spatially defined 'suburban trap' for middle-class women described by feminist geographers in the 1970s (Markusen, 1980) is disappearing, to be replaced by a *temporally* defined treadmill. Increasingly for suburban women, the problems of daily life are expressed in terms of time, and the solutions involve trading off time against other resources. The dream of a single-family suburban home is possible for most families now only through the full-time employment of both parents, and the working day is lengthened still further by the long commutes required from outlying affordable suburbs. Some husbands and wives turn to complementary work shifts to ensure that their children are cared for by at least one parent. At home, parents resort to a speed-up in daily life (Hochschild, 1989), rushing the children through bathing and eating and extending bedtimes to produce some 'quality time' together. Given these constraints on available time, alternatives such as employment in local back offices (Nelson, 1986) or working at home are welcomed, even though they carry costs of dead-end career trajectories.

What kind of new residential spaces are replacing the suburb, given the new kinds of household? Although demand for traditional single-family suburban housing continues, some other residential choices derived from women's employment are expressed in particular residential spaces and, in more general terms, in the relationship between jobs and housing location. Two household types can be singled out: two-worker households without children (including gay couples) and families headed by a single, employed mother.

The residential location choices of two-worker households without children will be affected by their generally higher disposable income, a different set of consumption patterns (such as a priority on restaurants rather than schools) and need to balance the commuting trips of two workers. However, the importance of the latter should not be overestimated; research by Hanson and Pratt (1988) found that contrary to the assumption in much economic theory that workplace locations are given and residential locations follow, for 93 percent of the employed women in their survey (and 66 percent of men), residential location preceded job search. Madden (1980, cited in Hanson and Pratt, 1988) found that households without children had significantly different location decisions from those with children and that single-person households were more likely to locate closer to their place of work than were other households. The two-worker childless households share many consumption patterns with single households, but their higher household income and greater spatial permanence means that they are more likely to be home owners, and hence gentrifiers, in some cities.

Female-headed families are more likely to live in urban areas than elsewhere and are more concentrated in central cities of metropolitan areas than the rest of the population (Mulroy, 1988). Their residential location is constrained by their low income and need for urban services such as public transport and accessibility to services such as child care. Female heads of households (including singles) make up over 40 percent of all the households living in rental units that have been defined as problem-ridden due to poor quality, overcrowding or excessive rent (Mulroy, 1988). And among these, it is single mothers who suffer the greatest problems with their housing. Winchester (1990) has identified parts of Australian cities where up to 25 percent of the population consists of lone-parent families. In the U.S., inner-city areas with public housing projects have high rates of single mothers. But although these inner-city zones may be notable for the high *concentration* of single mothers, Mulroy (1988) points out that about half of all female-headed families live in the suburbs, and their number is growing rapidly: it increased by 71 percent between 1970 and 1979, compared to a 41 percent growth rate in the central cities.

Although each of these three household forms (suburban families, two-worker households without children, and single mothers) is associated with distinctive urban spaces, they also compete for space. Single mothers in the suburbs tend to be viewed with suspicion, since they appear to be sliding down the 'ladder of life' associated with regular progression through stages of the family life cycle and housing tenure (Mulroy, 1988). Although divorce and separation usually displace families from suburban locations, when the family home is retained as part of the divorce settlement, mothers often attempt to develop new living and working arrangements to increase their income, such as subletting, sharing accommodations and running businesses from their homes. But rigid land-use regulations restrict the scope of these activities and often lead to the displacement of these 'nouveau poor' households.

Ironically, the expansion of women's employment has meant that the polarization of household income structure has increased and sometimes is expressed in conflict over the same urban space. For example, the central city redevelopment plans of Los Angeles are dedicated to expanding luxury accommodation for people who work in the highly-paid producer-services sector concentrated downtown (notably, childless professional couples and singles) to produce a '24-hour city.' Meanwhile, the people displaced from cheap housing in the inner city by redevelopment are often poor single mothers who work as office cleaners, retail clerks, seamstresses in garment sweatshops and routine clerical workers—jobs that are also clustered in downtown.

While the scenario described above highlights one potential spatial outcome, we also need to be aware of the way in which different conditions apply in each city and neighborhood. Rose and Villeneuve (1988), in a finely detailed study of inner-city Montréal, show how the changing employment structure (comprising a bipolarization and feminization of the labor force) coincided with shifts in the residential structure of inner-city neighborhoods. But instead of the clearly defined class and household distinctions identified above, they found a more complex mix, with, for example, single mothers in relatively low-paying professional jobs. They hypothesize that when gentrification involves the growth of this kind of population, the child-oriented motivations of these gentrifiers may potentially improve public services for all resident classes. As these single working mothers come to make up a significant share of the population in some areas, they may contribute to the creation of 'new' urban spaces of social reproduction.

The domestic division of labor

As more women have been to work outside the home and as more single-person and single-parent households have been established, the old domestic division of labor predicated on a full-time housewife has been called into question. Many more employed people now live in a household without a full-time housekeeper and must find new ways to integrate paid labor and domestic work. Hence the definition of necessary domestic work (and the form in which needs are met), the source of provision (by the market, the state, community or household) and the allocation of work within the household are all now subject to renegotiation.

One strategy for redefining work is through some reduction in standards. In families with two employed parents, the total number of hours spent on housework by husband and wife combined is usually less than when the wife is not in the labor force (Hochschild, 1989). Technological innovations have removed some of the manual labor involved in housework, but these reductions are all too often then matched by new demands, and much of the work involved in managing a household remains beyond the reach of technology.

A significant impact on domestic labor has resulted from the increase in market provision of services, ranging from new commodities that reduce labor (e.g., microwave ovens and prepared dinners) to more extensive use of services once used mainly by the wealthy (e.g., eating out in restaurants, house-cleaning, laundry and catering services). As more people engage in the sale of their labor power, more also engage in the purchase of goods and services. The extension of commodity relations of production into people's lives proceeds both through the constitution of individuals as producers of a commodity (wage labor) and as consumers of commodities. But access to market goods and services is sharply restricted by income, and this serves to reinforce class privilege. For example, when mothers of young children in both working-class and middle-class families routinely stayed home to care for them, the poverty of a home could be partly compensated for by the quality of non-material care (attention, warmth, etc.) provided by a mother, which was not necessarily determined by her class status and material resources. Now, the quality of care received by a child of a working mother is fairly directly determined by the amount that she is able to afford to pay for child care. When work such as child care is commodified, quality is related to price, and the reproduction of class privilege is intensified.

Access to market provision of domestic goods also reinforces gender privilege, given the lower average wage earned by women. So single women, who on average earn less than men of equivalent education and experience, will spend more time on housework: 886 hours per year for single women compared to 468 for single men, according to one study (Duncan and Morgan, cited in Hochschild, 1989). Single mothers, who bear the burden of both low income and heavy domestic responsibilities, are in the most difficult position.

The extension of market provision also has significant spatial and temporal implications for the role of the home. Not only are more items such as meals prepared outside the home, but they are also consumed outside: in a car, a restaurant, at a desk. The cluster of activities once firmly associated with 'home' begins to drift apart, and the blurring of boundaries between sites of home and work increases.

State services have not replaced services once performed in the home, although pressure is increasingly being placed on local governments for programs such as afterschool child care, health care in schools, and day care for dependent elderly. The role of the state is increasingly as a regulator and controller through legislation that presses other agencies into service, rather than as direct provider (Wolch, 1990). In this context, employers are increasingly being called on to fill the needs. Some employers have responded by offering fringe benefits such as on-site child-care centers, but these firms make up only a tiny fraction of all workplaces. Given the high cost of providing these services, they are likely to be offered to only the most valued members of the work force, whose company loyalty and increased productivity justify the expenditure.[6] The outcome is an intensified distinction between segments of the labor force, between the privileged group of permanent workers who receive these kinds of benefits and the workers with more tenuous attachments who are excluded.

[6]It is also questionable whether employers will regard services such as child care as the most effective means of attracting and retaining desired workers, given the continued inequality between men and women in terms of responsibility for child care and the consequent variation in the degree to which they value company-provided child care as a benefit.

Most of the daily work of social reproduction continues to be done by household members themselves, and with more family members working, there is less time to do these tasks. In this situation, demand has built up within households for ways to ease the pressure of time, through practices such as extended shopping hours, more flexible working hours and shorter commutes. But the intensity of time pressure is not felt equally by rich and poor or by men and women. People with lower incomes who are less able to purchase time-saving alternatives also suffer more deeply from the intensified pace of daily life. And despite the equalization in the hours worked by men and women outside the home, there has been very little renegotiation of the division of labor within the home. Children are now doing more food shopping and cooking, as well as spending more time alone at home in 'self-care.' But the bulk of the domestic work continues to be done by women even in households where both partners are employed for the same number of hours, resulting in a 'leisure gap' of about one month a year between men and women (Hochschild, 1989). One study of the hours spent on housework per year found that married women worked for 1,473 hours per year, compared to only 301 for married men (less than single men) (Duncan and Morgan, cited in Hochschild, 1989). This unequal division of labor means that the coupling constraints[7] and pressures on use of time are greater than they would need to be if work were more evenly shared.

The chief impact of these trends on the provision of services in the city has been to broaden both the spatial and temporal range of alternatives. More facilities such as supermarkets are open longer hours, and facilities such as restaurants are available both near the workplace and within residential areas, while apartment complexes and office buildings offer integrated facilities (gyms, laundries, etc.). But there is also a class difference involved in the *form* in which the demand for consumer services is met. For people at the lower end of the spectrum, the form in which commodification of household and personal activities takes place is most commonly through the purchase of a manufactured mass-market item, a do-it-yourself product.[8] For those at the upper levels of the income bracket, the essential appeal of the service is the made-to-order quality, which translates into the purchase of goods and services that are provided through some direct transaction between the provider and consumer and are thus more likely to involve the presence of the worker at the site of consumption. Accordingly, we see a polarization

[7] Coupling constraints refer to the possible preclusion of activities because of an individual's participation in other projects or because other individuals involved are unavailable. For instance, if an individual has two tasks separated by a three-hour drive and each task station is open for the same two-hour period, only one task can be accomplished on a given day. Similarly, in order to meet with a group of people, one must coordinate a time and place that is convenient for all.

[8] This may mean that an item takes the form of a manufactured good, which would be the case for a cooked meal in the form of a microwave-ready frozen dinner. However, a cooked meal prepared by a caterer would be classified as a service.

of consumption patterns, in line with the polarization of family and household incomes. And the labor-intensive demands of the higher-income consumption style generates an expanded supply of low-income service jobs, as Sassen shows in her study of New York (1984a, 1984b).

The effect is that spaces and times of personal fulfillment, entertainment and pleasure for the affluent are also spaces of work (exploitation) for those in labor-intensive service jobs. Evenings, weekends and public holidays are times of work for more and more people, as are public sites of entertainment, restaurants, shopping malls and resorts. And more and more, the requirement of that labor is that the act of labor (work-for-wage) be presented as an act of care (work-for-love). The requirement is not only that workers perform the *task* but that they accompany it with the *emotion*: the sincere smile, the warm greeting, the demand for what Hochschild (1983) refers to as 'emotional labor.' Thus elements of personal life and 'home' (true affection, sincerity, personal commitment) become colonized and blurred with 'work' in the irreducible site of power: the body and mind of the worker.

Flexibility in employment

It has been estimated that about one quarter of all jobs in the U.S. today are 'flexible'; that is, they involve some form of temporary, part-time or independent subcontracting employment relations (Christopherson, 1989). These alternative forms of attachment to the employer (or employment relations, for short) have some specific implications for the use of space in the city. Perhaps more important, they have changed the temporal patterning of urban activities, since many of the contingent employment relations involve a different daily schedule than is the case in traditional full-time, permanent employment.

The new forms of employment relation are in part produced by the drive to minimize wages and benefit costs, shift risk 'downwards' and maximize surplus value extraction. But the particular form that these impulses take is also shaped by the nature of the labor pool—in particular, by the set of constraints and opportunities which face different social groupings in the labor force. So, for example, the coincidence of low wages, limited promotion prospects, part-time hours and a largely female work force in some industries is not fortuitous: as Beechey and Perkins (1987) show in Britain, both the organization of work to produce part-time jobs and the conditions of those jobs are predicated on the existence of an accessible labor pool of married women and the prevailing domestic division of labor.

By emphasizing the contribution of workers to the shaping of working conditions, we do not mean to imply that the recent upsurge in contingent employment relations is a victory for workers. In fact, the proportion of female part-time workers who would rather have a full-time job has increased 300 percent from 1967 to 1985 (Appelbaum, 1987). Conversely, among full-time workers there are many (such as single mothers) whose heavy household responsibilities are such that they would rather work part-time, if only they could earn enough that way to survive. But any understanding of the prospects for change in working hours must take into account the real temporal constraints imposed upon people today by the need to balance work and family life and the choices they will make given a limited set of options.

Working hours have been changed in two ways by the extension of flexible employment relations. First, the distinction between times of work and times of personal life has been eroded on a societal as well as on a personal level. And second, this has been accompanied by a speedup and intensification of activity during the hours of work. So, although the boundaries between 'home' and 'work' are blurring, this has taken place in one direction only: work has intruded into domestic life.

The impact of service-sector expansion and the growth in paid employment by women on the societal experience of work time has already been mentioned. Entertainment, shopping, eating and other personal services all involve activities for which the timing is set by the consumer. And those consumers are increasingly demanding a wider range of time choices as they struggle to fit together the temporal demands of their jobs and household responsibilities. The growth in producer services has also contributed to an extension of work time, as the global integration of the financial system linked activities (such as stock markets) in different time zones. One outcome has been a more varied range of work schedules, leading to different commuting patterns.

The expression of flexible work schedules is experienced differently by people with different forms of attachment to an employer, but in each a process of intensification of work is apparent. Three typical forms of attachment can be illustrated: part-time worker, full-time worker, and independent contractor.

Part-time work in some cases is a way of reorganizing work so that only the times of most intense labor demand are paid for; the rest of the time is 'free.' In the U.S. today, over 71 percent of part-time jobs are in wholesale or retail trade or in services (Appelbaum, 1987), sectors where the labor demand is subject to seasonal and diurnal peaks. And some of the paid rest periods built into the eight-hour day can also be avoided; for example, work may be broken into six-hour shifts that avoid the requirement of a paid 'unproductive' lunch break. Part-time work can also be a strategy to increase productivity through more intensive labor, since many part-time jobs could not be maintained at the same level of intensity for an eight-hour day (Beechey and Perkins, 1987).

The growth in part-time work will affect the time-space activity patterns of workers, although the direction of change is complex and may consist of contradictory trends. For example, if part-time work is in the form of a few hours each day, it will tend to attract workers who live close by (in other words, draw on a smaller spatial labor market than full-time work), since the travel time and costs will absorb a greater share of the wages than full-time work. However, if part-time means a few long shifts each week, this makes feasible a longer commuting distance between home and work, and the spatial labor market would be potentially greater. On the face of it, the second work schedule seems to impose lower direct costs on workers, yet this depends on their ability to reschedule other tasks, a matter of control over their time which is related to gender and family situation. European studies (cited in Appelbaum, 1987) have found that, given the choice of how to reduce working hours, men tend to prefer gaining free time in a few large blocks, such as an extended weekend, while women would prefer a shorter working day. But part-time hours are not necessarily stable; increasingly, part-time work is being combined with flexible work schedules. For example, workers such as airline attendants may be 'on call' 24 hours a day, yet only work a few hours a month. Also, part-time work does not necessarily mean a shorter working week; in many cases part-time jobs supplement other full-time or part-time jobs. Multiple job-holding has declined among men (from 7 percent in 1970 to 5.8 percent in 1980), but it has risen by women, from 2.2 percent to 3.8 percent in 1980 (Appelbaum, 1987).

For full-time workers (and here we include temporary workers with full-time schedules), the intensification of work has taken place not by controlling working hours, but by defining responsibility in terms of completing tasks rather than putting in the hours. The protections of hourly workers are being lost as working conditions are redefined with reference to the standards governing professional work practices (but without all the benefits of professional jobs). Examples include the team system in the auto industry and changing practices in retail sales. For example, in Nordstrom Department stores, sales clerks have been required to write thank-you notes and make deliveries after they have clocked out for the day. The extension of the average working day without compensation is an integral component in the latest strategy to maximize profits. This also affects those professional and middle-management workers who have historically been expected to work occasionally without overtime pay but who now find those demands solidified. For example, the 'billable hours' expected to be generated by lawyers in corporate firms has risen in recent years, requiring longer hours in the office. Technologies such as lap-top computers, cellular phones and modem links from a home computer are all used to ensure that even time spent travelling or commuting is also work time. So, for example, the typical day of one middle-management hospital administrator begins at 7 a.m. at her home computer, lasts from 8 a.m. to 7 p.m. at the office and stretches into evening and weekends as well (Libman, 1988). The effects of these trends are shown in the figures for leisure time in the U.S., which has declined over the past decade.[9]

The speedup of middle management daily work is also contributing to a gendered distinction, or at least a distinction between those men and women without children who can put in a ten-hour working day and those (usually women) with primary responsibility for the care of the children. One outcome is a 'streaming' process, either explicit as in a 'mommy track' or covert in terms of the hidden benefits that accrue to those with backstage domestic support in the form of a non-employed spouse.

Given the more stringent demands of many full-time jobs in large firms, self-employment seems to offer some a greater degree of control over the scheduling of their working day, an attractive possibility to many women with children. This impetus coincides with the drive toward sub-contracting described above and is supported by an ideology of entrepreneurship which conceals the subordinate relationship of most to the agents they serve. Daily work schedules may vary, depending on the amount of work and deadlines to be met. But flexibility in time is matched by flexibility in space: a common element of these small businesses, especially newly-formed firms, is that the primary workspace is located in the home.

In this brief survey, it is clear that a wider range of potential daily work schedules is now possible. While this may facilitate some households in meeting daily needs (such as parents who work complementary shifts), it also places some serious strains on individuals as they attempt to match their schedules to those of child-care facilities, shops, clinics, and so on. The demand for more flexible conditions of employment—which allow some say over the times of work, as well as some variety from the traditional eight-hour day—reflects those needs. For example, although temporary work is often promoted for its variety and interest, one study of temporary workers found that over 60 percent said the freedom to schedule work in a flexible manner was the most important factor in their choice of temporary work; only 17 percent said that variety was the reason (Appelbaum, 1987).

The impact of flexible work patterns on urban *space* is primarily through the transformation of the home into a site of work. As with the incursion into the personal time of the worker, the form of this incursion varies by employment relation and by gender. For middle management, work has long been brought home. One study of AT&T workers in 1982 found that over 30 percent took work home (Appelbaum, 1987), and the extension of communications technology such as fax machines and on-line computer linkages has made possible a form of telecommuting for executives. But for

[9]A Harris Poll found that the average workweek increased in the U.S. from 40.6 hours in 1973 to 46.8 hours in 1987, and during the same period, leisure time shrank to 16.2 hours per week, down from 26.2 hours. Professional people worked an average of 52.2 hours a week, while those in small businesses worked for 57.3 hours (Libman, 1988)

workers at lower skill levels, the new technology has generated a form of homework not very different to traditional forms, even as the traditional forms of homework in the garment, jewelry and other craft industries are experiencing a revival (Fernandez-Kelly and Garcia, 1985).

The number of workers and firms participating in fairly large-scale, regular arrangements of electronic homework is still relatively small, but the prospect is receiving much interest. A recent estimate is that 250 firms are involved, with 10,000 employed at home using computer technologies (Appelbaum, 1987). The arrangement offers significant savings for firms: workers receive less of the standard employment benefits, especially health insurance, they must rent equipment and pay for the cost of overheads such as rent and heating and they are often paid by piece rates which make no allowance for time spent setting up, delivering material and dealing with problems. Sometimes this is also a way for companies to make use of computer facilities during off-peak times.

More traditional homeworking is also on the upswing, both in legal and irregular forms. During the Reagan administration, restriction of homeworking in the production of jewelry, sweaters and other products was relaxed. While the restrictions on homework in garment production itself have been maintained, the practice is widespread. The *Los Angeles Times* (Efron, 1989) reported in 1989 the case of a woman and her seven-year-old child who had been sewing together garments at home, earning $1.45 per hour. But while flexible working practices transform the home into a site of work even for small children, the worksite is less and less home-like, with speedups in every field: ever-tighter control of social interaction between workers and customers through the enforcement of standardized scripts (Garson, 1988) and rearrangement of working hours into shorter shifts that, by excluding paid breaks, also attack the possibility of non-work-related interaction.

While the conflict over the extension of flexible employment conditions is explicable in terms of the conflict between capital and labor, the same terrain is also the site of struggle over, for example, the way that gender relations are reproduced into patterns of power, access to resources, and so on. The outcomes are complex. So for many people, flexible working conditions represent a prize in the struggle to construct a livable life, even while they represent a failure in the class struggle or a failure in the 'stalled revolution' (Hochschild, 1989) in gender relations over sharing domestic labor in the home.

Conclusions

Although restructuring is expressed differently in the institutions of the economy, the state, the community and the household, some broad themes can be identified: fragmentation of monolithic social institutions and their replacement by a multiplicity of alternatives rather than a new monolith and a blurring of the boundaries between times and spaces that were sharply delineated in the preceding period. So, for example, the numerical dominance of the employed-father/housewife-mother nuclear family household has been replaced by a diversity of households, just as the dominance of the full-time, year-round job in a physically discrete factory or office has been replaced by a diversity of employment relations. And this has been accompanied by a diversity of social practices that blur the boundary between, for example, what constitutes a worksite (e.g., factory), working shift (e.g., 'office hours') and work as an activity (e.g., making an object) as distinct from the sites, times and activities that made up the home and personal life. In the emerging system, the site of employment is now more likely than before to be variable (such as the end of a telephone or modem), working hours are more likely to be irregular and the employment activity is more likely to involve some form of emotional labor.

Of course, these themes do not simply represent changes in the way the world *is*; they are perhaps more indicative of changes in the way the world is perceived, both in common sense and in theoretical terms. Diversity and blurred boundaries result from the erosion of an imposed social consensus and the 'discovery' of long-neglected and marginalized social practices as much as they represent entirely new occurrences. This double transformation—in events and in perceptions—is most clearly expressed in the writings on post-modernism. The explicit linkage between the conditions of post-modernity (experiences of fragmentation and discontinuity arising from social and economic trends) and the theoretical approach of post-modernism (attention to a multiplicity of viewpoints, to the repressed voices and silences in any discourse) has served to simultaneously reveal and validate previously marginalized social institutions and actors.

In this new awareness of diversity, a richer understanding of social reproduction is possible, and this is expressed in some recent work. With regard to wage labor, attention is being focused on the employment conditions (such as casual labor) once thought to be mere relics of a past age. The deindustrialization of cities—specifically, the loss of the large factory-based work opportunities that provided the symbolic and to a large degree actual foundation of working class survival in the city—has renewed attention to the informal sector and the way in which it is integrated into both the formal economy and the household economy (Redclift and Mingione, 1985). The shrinking of the role of the welfare state as a direct supplier of services has turned attention to the ways in which such services are provided by the voluntary sector (Wolch, 1990) and the variety of ways in which the state continues to intervene in the provision of these services by households and communities. As Rose (1990) points out, social needs such as child care have long been met by arrangements such as paid baby-sitting by neighbors, but the focus on collective consumption issues to the neglect of non-state forms of provisioning meant that these were understudied.

A revived interest in the informal sector corresponds to a renewed awareness of the changing nature of the social division of labor. There is a sharper perception that social reproduction in most households is achieved through patching together a variety of resources. Also, there is a more widespread recognition that the form in which those resources are provided is historically and spatially variable and subject to renegotiation. And more attention is being paid to the ways in which the provision of resources by each agency interacts with the other. The labor market, state, community and household are not only agencies of provision but also structured environments that differentially privilege various agents. Since these sources of support do not operate in isolation, the form and level of support available in one sphere enables or constrains people in drawing on other sources. So, for example, as Fincher (1989) has pointed out with reference to the local state, the level and form of provision of services such as child care profoundly affects the ability of women to participate in the labor market.

In the context of an historically specific division of labor, the concept of social reproduction provides a valuable point of entry to understanding the linkages between a range of activities in seemingly separate realms, including wage labor, the welfare state, community and household. Each of these represents alternative and complementary routes through which individuals make claims upon others to meet their needs for daily survival. These also represent alternative structures within which political identity is constituted. So, for example, family role or ethnicity are components of subjective identity that are simultaneously socially constructed and individually manipulated as resources and that are potential poles around which political activism can be mobilized. But the nature of these structures (labor market, state, community and household) is also variable. The social division of labor within which strategies of social reproduction are carried out is constantly subject to renegotiation, and the direction of transformation has diverging implications for people in different social positions. Given the depth of the transformations that may be observed, it is perhaps opportune to question the persistence of the categories of analytic thought in academic discourse that separate 'production' so distinctly from 'reproduction.'

Acknowledgments

We would like to thank Lois Takahashi for her helpful comments on an earlier version of this chapter. The support of the National Science Foundation (SES 89-21241) is also acknowledged.

References

Aglietta, M. 1979 *A Theory of Capitalist Regulation.* London: Verso.

Alliez E. and **M. Feher.** 1987 The Luster of Capital, *Zone,* 1:2, 314–59.

Appelbaum, E. 1987 Restructuring Work: Temporary, Part-time and At-Home Employment. In Hartmann, Heidi (ed.), *Computer Chips and Paper Clips: Technology and Women's Employment.* Washington, D.C.: National Academy Press.

Baran, B. 1985 Office Automation and Women's Work: The Technological Transformation of the Insurance Industry.' In M. Castells (ed), *High Technology, Society and Space.* Beverly Hills: Sage Publications.

Beechey, V. and **T. Perkins.** 1987 *A Matter of Hours: Women, Part-Time Work and the Labor Market.* Minneapolis: University of Minnesota Press.

Bluestone, B., W. R. Murphy, and **M. A. Stevenson.** 1973 *Low Wages and the Working Poor.* Ann Arbor, Mich.: The Institute of Labor and Industrial Relations, University of Michigan.

Brooks, K. and **M. Ness.** 1990 Jobless-Insurance Cuts: Out of Work? Out of Luck. *The Nation,* vol. 251, no. 22 (December 24, 1990), pp. 800–804.

Chavez, S. and **J. Quinn.** 1987 Garages: Immigrants In, Cars Out, *Los Angeles Times* A1, 24 May 1987.

Chen, Yung-Ping. 1982 Changing Family Roles: Their Impact on Benefit Programs. In D.I Salisbury (ed.), *America in Transition: Implications for Employee Benefits.* Washington, D.C.: Employee Benefit Research Institute.

Christopherson, S. 1989 Labor Flexibility in the United States Service Economy and the Emerging Spatial Division of Labor, *Transactions, Institute of British Geographers,* 14, 131–43.

Cohen, R.B. 1981 The New International Division of Labour, Multinational Corporations and Urban Hierarchy. In A. J. Scott and M. Dear (eds.), *Urbanization and Planning in Capitalist Societies.* London and New York: Methuen.

Daniels, A. K. 1986 *Invisible Careers: Women Community Leaders in the Voluntary World.* Chicago: University of Chicago Press.

Danziger, S. and **D. Feaster.** 1985 Income Transfers and Poverty in the 1980s. In J. M. Quigly and D. L. Rubinfeld (eds.), *American Domestic Priorities: An Economic Appraisal.* Berkeley: University of California Press.

Davis, M. 1990 *City of Quartz.* London: Verso Books.

Dear, M. J. and **J. R. Wolch.** 1987 *Landscapes of Despair: From Deinstitutionalization to Homelessness.* Princeton, NJ: Princeton University Press.

Doeringer, P. and **M. Piore.** 1971 *Internal Labor Markets and Manpower Analysis.* Lexington, Mass.: Heath Books.

Efron, S. 1989 Mother's Plight Turns a Home to Sweatshop. *Los Angeles Times* Nov. 27, 1989. Part I.

Ehrenreich, B. 1983 *The Hearts of Men: American Dreams and the Flight from Commitment.* Garden City, NY: Anchor Press/Doubleday.

Ellwood, D. T. 1986 The Spatial Mismatch Hypothesis: Are There Teenage Jobs Missing in the Ghetto? In R. Freeman and H. Holzer (eds.), *The Black Youth Employment Crisis.* Chicago: University of Chicago Press, pp. 147–90.

Fernandez-Kelly, M. P. and **A. Garcia.** 1985 The Making of an Underground Economy: Hispanic Women, Home Work, and the Advanced Capitalist State, *Urban Anthropology* 14, 1–3.

Fincher, R. 1989 Class and Gender Relations in the Local Labor Market. In J. Wolch and M. Dear (eds.), *The Power of Geography*, pp. 91–115. Boston: Unwin Hyman.

Frobel F., J. Heinrichs, and **O. Kreye.** 1980 *The New International Division of Labour: Structural Unemployment in Industrialized Countries and Industrialization in Developing Countries.* New York: Cambridge University Press.

Garson, B. 1988 *The Electronic Sweatshop: How Computers are Transforming the Office.* New York: Simon and Schuster.

Gilbert, N. 1983 *Capitalism and the Welfare State: Dilemmas of Social Benevolence.* New Haven: Yale University Press.

Giloth, R. and **J. Betancur.** 1988 Where Downtown Meets Neighborhood: Industrial Displacement in Chicago, 1978–1987, *Journal of the American Planning Association*, 54, 279–90.

Gleeson, B. R. and **J. R. Wolch.** 1989 Homelessness and the Politics of Turf: The Case of Venice, California, Working Paper 17, Los Angeles Homelessness Project. Los Angeles: Department of Geography, University of Southern California.

Gordon, D., R. Edwards, and **M. Reich.** 1982 *Segmented Work, Divided Workers.* Cambridge: Cambridge University Press.

Gordon, D., R. Evans, and **M. Reich.** 1975 *Labor Market Segmentation.* New York: D.C. Heath Co.

Gordon, P., H. W. Richardson, and **H. L. Wong.** 1986 The Distribution of Population and Employment in a Polycentric City: The Case of Los Angeles, *Environment and Planning, A* 18, 161–73.

Hanson, S. and **G. Pratt.** 1988 Reconceptualizing the Links Between Home and Work in Urban Geography: Review, Critique, Agenda, *Economic Geography*, 64, 299–321.

Harrison, B. and **B. Bluestone.** 1988 *The Great U-Turn.* New York: Basic Books.

Hartmann, H. 1987 Changes in Women's Economic and Family Roles in Post-World War II United States. In L. Beneria and C. R. Stimpson (eds.), *Women, Households and the Economy.* New Brunswick: Rutgers University Press.

Hochschild, A. R. 1983 *The Managed Heart: Commercialization of Human Feeling.* Berkeley: University of California Press.

Hochschild, A. R. with **A. Machung.** 1989 *The Second Shift: Working Parents and the Revolution at Home.* New York: Viking.

Holloway, S. R. 1990 Urban Economic Structure and the Urban Underclass: An Examination of Two Problematic Social Phenomena, *Urban Geography*, 11, 319–46.

Hopper, K. 1990 Poverty and Homelessness. Paper presented at *Broadening Perspectives on Homelessness: An Interdisciplinary Symposium* Los Angeles, December 7, 1990.

Hughes, M. A. 1989 Misspeaking Truth to Power: A Geographical Perspective on the 'Underclass' Fallacy, *Economic Geography*, 65, 187–207.

Husick, T. and **J. R. Wolch.** 1990 On the Edge? An Analysis of Homed and Homeless Applicants for General Relief in Los Angeles County, Working Paper 29, Los Angeles Homelessness Project. Los Angeles: Department of Geography, University of Southern California.

Johnson, J. H. and **C. C. Roseman.** 1990 Recent Black Outmigration from Los Angeles: the Role of Household Dynamics and Kinship Systems, *Annals of the Association of American Geographers*, 80, 205–22.

Kain, J. F. 1987 Housing Market Discrimination and Black Suburbanization in the 1980s. In G. Tobin (ed.), *Divided Neighbourhoods.* Beverly Hills: Sage Publica tions.

Kasarda, J. D. 1989 Urban Industrial Transition and The Underclass, *The Annals of the American Academy of Political and Social Science*, vol. 501 (edited by William Julius Wilson).

Katz, M. B. 1989 *The Undeserving Poor: From the War on Poverty to the War on Welfare.* New York: Pantheon Books.

Kuhn, S. and **B. Bluestone.** 1987 Economic Restructuring and the Female Labor Market: The Impact of Industrial Change on Women. In Lourdes Beneria and C. R. Stimpson (eds.), *Women, Households and the Economy.* New Brunswick: Rutgers University Press.

Law, R. and **J. R. Wolch.** 1991 Homelessness and Economic Restructuring, *Urban Geography*, 12, 105–36.

Law, R. M., J. R. Wolch and **L. Takahashi.** 1991 The Future of Technopolis: Militarized Industrial Spaces in Southern California, Working Paper 31, Los Angeles Homelessness Project. Los Angeles: Dept. of Geography, University of Southern California.

Levy, F. 1988 Incomes, Families and Living Standards. In R. E. Litan, R. Z. Lawrence, C. L. Schultze (eds.), *American Living Standards: Threats and Challenges.* Washington, D.C.: The Brookings Institute.

Libman, J. 1988. Why We Overwork, *Los Angeles Times*, June 13, 1988, Part V.

Lipietz, A. 1986 New Tendencies in the International Division of Labor: Regimes of Accumulation and Modes of Regulation. In M. Storper and A. Scott (eds.), *Production, Work and Territory.* London: Allen and Unwin.

Markusen, A. 1980 City Spatial Structure, Women's Household Work and National Urban Policy, *Signs*, 5:3.

Mattera, P. 1990 *Prosperity Lost.* Reading, MA: Addison-Wesley, Inc.

Mulroy, E. (ed.) 1988 *Women as Single Parents: Confronting Institutional Barriers in the Courts, the Workplace, and the Housing Market.* Dover, Mass: Auburn House Publishing Co.

Nelson, K. 1986 Labor Demand, Labor Supply, and the Suburbanization of Low-wage Office Work. In M. Storper and A. J. Scott (eds.), *Production, Work and Territory.* London: Allen & Unwin.

Noyelle, T. 1987 *Beyond Industrial Dualism: Market and Job Segmentation in the New Economy.* Boulder, CO: Westview Press.

Palmer, J. L. and **I. V. Sawhill** (eds.) 1982 *The Reagan Experiment: An Examination of Economic and Social Policies under the Reagan Administration.* Washington, D.C.: The Urban Institute Press.

Redclift, N. and E. Mingione (eds.) 1985 *Beyond Employment: Household, Gender and Subsistence.* London: Basil Blackwell.

Rose, D. 1990 Collective Consumption Revisited: Analysing Modes of Provision and Access to Childcare Services in Montreal, Quebec, *Political Geography*, 9, 353–80.

Rose, D. and **P. Villeneuve.** 1988 Women Workers and the Inner City: Some Implications of Labor Force Restructuring in Montreal, 1971–1981. In C. Andrew and B. Moore Milroy (eds.), *Life Spaces: Gender, Household, Employment.* Vancouver: University of British Columbia Press.

Roth, R. 1990 Perceptions of Menace: The Homeless and the Downtown Economy, *Pacific Northwest Executive*, 6:3.

Rowe, S. and **J. R. Wolch.** 1990 Social Networks in Time and Space: Homeless Women in Skid Row, Los Angeles, *Annals of the Association of American Geographers*, 80, 184–205.

Saks, D. H. 1983 *Distressed Workers in the Eighties.* Washington, D.C.: National Planning Association.

Sassen, S. 1984a The New Labor Demand in Global Cities. In M. P. Smith (ed.), *Cities in Transformation.* Beverly Hills: Sage Publications.

Sassen, S. 1984b Growth and Informalization at the Core: A Preliminary Report on New York City. Paper presented at the Symposium on the Informal Sector, Johns Hopkins University.

Sassen, S. 1988 *The Mobility of Labor and Capital*, Cambridge: Cambridge University Press.

Sawhill, I. V. 1988 Poverty in the U.S.: Why is it so Persistent? *Journal of Economic Literature*, Vol. XXVI (September), 1073–119.

Scott, A. J. 1984 Industrial Organization and the Logic of Intra-metropolitan Location, III: A Case Study of the Women's Dress Industry in the Greater Angeles Region, *Economic Geography*, 60, 1–27.

Shilts, R. 1987 *And the Band Played On: Politics, People and the AIDS Epidemic.* New York: St. Martin's Press.

Sidel, R. 1986 *Women and Children Last: The Plight of Poor Women in Affluent America.* New York: Penguin.

Smith, N. and **P. Williams** (eds.). 1986 *Gentrification of the City.* Boston: Unwin Hyman.

Smith, S. 1990 Southern Discomfort. In *These Times*, 15, No. 7, Dec. 26, 1990–Jan. 15, 1991, p. 12.

Wall Street Journal. 1990 Part-time Work: There's a Dark Side to All That Flexibility, November 27, 1990, Page A1.

Weiss, M. J. 1988 *The Clustering of America.* New York: Harper and Row.

Williams, B. 1988 *Upscaling Downtown: Stalled Gentrification in Washington, DC.* Ithaca, NY: Cornell University Press.

Wilson, W. J. 1987 *The Truly Disadvantaged: The Inner City, the Underclass, and Public Policy.* Chicago: University of Chicago Press.

Winchester, H. 1990 Women and Children Last: The Poverty and Marginalization of One-Parent families, *Transactions of the Institute of British Geographers*, 15, 70–86.

Wolch, J. R. 1990 *The Shadow State: Government and Voluntary Sector in Transition.* New York: The Foundation Center.

Wright, T. and C. Riave. 1989 The Decline of Orange County's Suburban Oasis: Regional Problems in Employment, Housing, and Transportation, *International Journal of Sociology and Social Policy*, 9, 97–117.

Eight

THE POSTMODERN URBAN MATRIX

Paul L. Knox

Virginia Polytechnic Institute and State University

The evolution of urban form has been characterized by a continuous restlessness, founded on a series of epochal transformations: from the mercantile city of the mid-nineteenth century through the transitional city of competitive industrialization to the organized city of corporate control and state management and, now, to the beginnings of a new, post-industrial, postmodern city. As emphasized in Chapter 1, the emergent form of the late-twentieth-century city must be understood in the context of the simultaneous existence within cities of both surplus labor and surplus capital (the overaccumulation crisis) and in the consequent redeployment of corporate investments and activities and the emergence of lean and flexible systems of industrial production. It must be understood in the context of the redrawing of the relationships between the public and private sectors, the increasing scale of corruption, the growth of materialism and the increasing intensity of socio-economic polarization. It must be understood in the context of the implosion of the ordered, Euclidian space of modernity and the emergence of a 'hyperspace' of ideas, information and images in which the autonomy of places is being undermined and their qualities diminished while the economic and socio-cultural significance of remaining differences between them is intensified. And it must be understood in the context of the influence of postmodernity in a broad spectrum of urban life. The matrix of settings and fragments that has emerged from this context must be seen, in turn, as the basis for a new socio-spatial dialectic from which the spatiality of the early twenty-first-century city will emerge. In this chapter, I unpack the material landscapes of one metropolitan area—Washington, D.C.—in order to illustrate the nature of these fragments and their relationship to the ecology of the city.

The Washington metropolitan area

At first glance, the Washington, D.C. metropolitan area may not seem a likely candidate for an exercise of this sort. Described by John Kennedy as the only city to combine northern charm with southern efficiency and by Richard Nixon as a city full of 'pointy-headed bureaucrats,' Washington has for a long time endured the reputation of being a rather dull federal town, lacking the dynamism of industrial centers and failing to capture the cosmopolitanism of major corporate and financial control centers. The city's material landscapes have reflected these shortcomings. The physical centerpiece of the city, L'Enfant's plan, had been allowed to leak purposelessly away in a series of compromises and missed opportunities, while the architecture of public and commercial buildings, under a height restriction of 110 feet imposed in 1910 in order to preserve the visual dominance of the Capitol and the Washington monument, made for a rather bland, stodgy and sterile urban core. In the past fifteen years, however, the Washington metropolitan area has been transformed from a federal town to a world city (Gale, 1987; Knox, 1987, 1991). Although government jobs—state, local and federal—still account for one in every four jobs in the metropolitan area, it is now a national information and communications node and a major coordinating center for all kinds of international activities, both public and private. This has resulted in a simultaneous recentralization of commercial activity and emergence of 'edge cities' (Garreau, 1991); and it has created a congested, fragmented and polarized urban structure that is a good example of the postmodern metropolis.

The largest sector of the metropolitan economy is the service sector. In part, this is a product of the tourism generated by monumental Washington and the complex of galleries and museums around the Mall; but more important are the office jobs generated by the interest groups and corporations attracted to Washington by the presence of the national legislature, its bureaucratic agencies and the bureaux of major international agencies such as the World Bank. Washington contains the headquarters of nearly 200 national business, professional and trade associations and a growing number of law, accounting, real estate, computer software and high technology firms, media and communications companies, data services, mortgage banks and investment trusts and consultants ('Beltway bandits') such as economists, management analysts and scientists who do contract work for both large corporations and government agencies. Foreign-owned firms such as British Aerospace, Arianespace, Fokker Aircraft, Polaris Optics and Lafarge concrete and construction have begun to locate their US headquarters in Washington,[1] which has also become the home of a growing number of Fortune 500

[1]Altogether, more than 130 foreign-owned companies have selected Washington as the headquarters of their U.S.

(continued...)

companies, including MCI in the District, Martin Marietta and Mariott in suburban Maryland and Gannett (owners of *USA TODAY*), Mobil, Unisys and US Sprint in suburban Virginia. At the same time, there has been a pronounced shift away from direct federal employment toward government purchasing. Between 1979 and 1989, federal purchases of local goods and services trebled, from $3.4 billion to $10.3 billion, reaching a level that approaches the dollar value of the area's federal payrolls.[2] Thousands of local companies now live by federal contracts, with the lion's share going to companies that provide communications equipment, defense systems R&D, social services, architectural and engineering services, data processing services and professional services.

Overall, the Washington area acquired almost 600,000 new jobs—one fourth of the current total—during the 1980s (Granat and Conlin, 1990). In the process, it became an 'entrepreneurial city,' with one of the highest rates of growth in the nation for new business enterprises with ten or more employees and the second-highest percentage of young companies with high growth rates.[3] The result of all this growth is an exceptionally well-educated and affluent population. One in three adults are college-educated, compared to one in five in metropolitan Chicago, New York and Los Angeles and one in seven in metropolitan Detroit. In 1990, when the national average household income was $28,525, the average for the Washington metropolitan area was $43,754. One in four of Washington's black households is affluent,[4] compared to one in ten in Detroit and Chicago and one in twenty-five in Miami (Waldrop, 1990). Washington is the strongest consumer market in the country, with average retail sales per household of $22,454 in 1988. Within the metropolitan area, expenditure on men's tailored clothing, precious jewelry, imported cars, foreign vacations, imported wine, dining out and health club membership runs at more than 25 percent above the U.S. average (Weissman, 1990).

Among those at the top end of the income distribution, physicians in the Washington area had incomes that averaged $130,000 in 1989—20 percent higher than the national average; partners in large law practices had incomes in 1989 in excess of $250,000, with some[5] averaging well over $500,000; and the senior officers of 130 of the largest

[1](...continued)
operations, about five times the 1980 number. Together these companies employ about 16,000 people (*Washington Post*, March 7, 1990, p. C1)

[2]Based on data from the Greater Washington Research Center.

[3] As reported by *Inc.* magazine. See J. Case, The most entrepreneurial cities in America, *Inc.*, March 1990, 41–48.

[4]With an income at least five times the poverty threshold in 1988.

[5]In the offices of Skadden, Arps, Slate, Meagher and Flam and Fried Frank, Harris Shriver and Jacobson. See Van Dyne (1989).

business and professional associations averaged almost $200,000 in 1987, with seventeen of them earning over $300,000 and four[6] earning over $500,000. It is estimated[7] that there are more than 277,000 households in the Washington area with net (after-tax) annual incomes of $75,000 or more, over 112,000 with net incomes of $100,000 or more, and more than 15,000 with net incomes of $250,000 or more—about three times the national incidence of such incomes.

The economic growth and rising affluence of the 1980s was reflected in Washington in a materialism that took its cue from the inauguration of Ronald Reagan as President in 1980:

> *The staff of the Washington Post's Style section was kept busy for days counting the private jets at the airport, measuring the length of limousine queues, inspecting the quality of mink garments, interviewing Reagan's millionaire friends from California, and inventorying the first lady's wardrobe. All of this was an early indication that the Reagan years would see a renewal of the idea that the making and spending of money is a noble goal and a moral good.*
>
> Van Dyne, 1989, p. 250

The ensuing development boom and consumption spree has been very visible in office, retailing and residential development. During the late 1980s, more office space was completed in the Washington area than in any other North American or European city.[8] In contrast to the plainness of Washington's federal office space, much of this office space is deliberately and self-consciously luxurious; some of it is downright outré. The executive suite of the Gannett Company's office tower in Rosslyn is decorated with snakeskin wallpaper, suede rugs, and two 'neorealist' sculptures of sheep, complete with small piles of Hershey's chocolate Kisses strategically located under their rears (Van Dyne, 1989). Retailers have been quick to exploit the area's buying power. In the late 1970s, Bloomingdale's, Nieman Marcus and I. Magnin department stores appeared, followed later by Nordstrom's, Macy's, Saks Fifth Avenue and Lord and Taylor, with Hecht's commissioning the only full-sized, freestanding department store to be built in a U.S. downtown since 1945. New shopping malls and gallerias have attracted upscale specialty stores such as Abercrombie and Fitch, Laura Ashley, Bally, Burberys, Cartier, Graham and Gunn, Gucci, Hermés, Ralph Lauren, Yves St Laurent, FAO Schwartz,

[6]The executive directors of the American Medical Association, the Motion Picture Association of America, the National Cable Television Association and the Edison Electric Institute.

[7]Claritas Corporation, Alexandria, Virginia; cited in Van Dyne (1989).

[8]*Washington Post*, February 11, p. E1.

Sharper Image, Tiffanys, Louis Vuitton and Williams-Sonoma; automobile dealerships have added Rolls Royces, Bentleys, Ferraris and Lamborghinis to the BMW, Jaguar, Mercedes Benz, Porsche and Volvo models on their lots.

The most striking outcome of Washington's development boom and consumer spree, however, is in the residential sector. The expansion of well-paid jobs brought large numbers of households in their prime home-buying years to the area, pushing up house prices and prompting a rash of construction for the top end of the market. According to real estate analysts R. S. Lusk and Son,[9] there were seventeen Washington-area neighborhoods where average home sale prices exceeded $350,000 in 1988 (Table 8.1). New homes in Potomac and Great Falls stand on large lots, with circular drives and imposing gateways that create the unnerving effect of a landscape full of expensive funeral homes (Fig. 8.1). The houses themselves average 6,000 to 10,000 square feet, though some are the size of a three-story public school. They have elaborate master bedrooms, marble bathrooms with Jacuzzis, saunas or steam cabinets, exercise rooms, 'gourmet' kitchens, libraries with computer centers, two-story foyers and 10-foot ceilings. Optional extras include cottages for guests, riding rings for horses and holding tanks for lobsters. When the boom that began in 1984 eventually slowed in 1990, such features, along with locational advantages, became critical in securing a competitive advantage in sluggish markets (Parham, 1990).

As in other world cities, economic growth and conspicuous consumption have been accompanied by an intensification of poverty and social malaise. The District of Columbia contains neighborhoods whose conditions are as shockingly bad as those in Potomac and Great Falls are ridiculously good. Their condition is less well-documented because there is less money to be made from them: for real estate analysts and marketing companies they are all but invisible. They consist of 2- and 3-story apartment buildings and some older single-family dwellings and row houses that have deteriorated to the point of no return. They are inhabited by a residual population that has been unable to participate in the expansion of the advanced service economy and so unable to escape to better neighborhoods.

Yet, while these neighborhoods are invisible to marketers and to the more affluent residents of the metropolitan area, they are the locus of two of the attributes by which Washington is best-known: homicide and drug abuse. Washington, D.C. has become firmly established as the nation's Murder Capital. With 487 homicides in 1991 (up from 456 in 1990, 438 in 1989, 369 in 1988, 228 in 1987 and 197 in 1986), the city had a homicide rate of 72 per 100,000 residents, almost exactly twice that of neighboring Baltimore. Ninety percent of homicide victims in the District were, like their assailants,

[9]Cited in Van Dyne, 1989, p. 248

black:[10] citizens of the city of the excluded, invisible until they showed up in a body bag on the six o'clock news. The link between crime and drugs is very close: two-thirds of homicide victims in recent years have had drugs in their systems,[11] and a similar proportion of defendants in the District's Superior Court have been found to be drug users.[12] Many of the homicides are the result of execution-style killings, casualties of the turf wars that have proliferated in the absence of organized crime and the presence, within the city of the excluded, of an over-supply of aggressive people seeking to escape poverty or to pay for their own habit by dealing drugs.

Table 8.1. The Most Expensive Washington-Area Neighborhoods, 1988

Neighborhood	*Average Price of Homes Sold in 1988*
Massachusetts Heights	$1,200,000
Kalorama	$808,000
Berkeley	$686,000
Spring Valley	$675,000
Forest Hills	$594,000
Woodley	$581,000
Wesley Heights	$576,000
Great Falls	$499,000
Kent	$488,000
Potomac	$478,000
Georgetown	$464,000
Observatory Circle	$448,000
McLean	$422,000
Burleith	$414,000
Cleveland Park	$387,000
Chevy Chase	$379,000
West Bethesda	$374,000

Source: Van Dyne, 1989, p. 248

[10]The Puzzle of DC's Deadly Distinction, *Washington Post*, October 30, 1989; pp. A8–9.

[11]Amid Endless Killings . . . , *Washington Post*, January 22, 1990, pp. E1, E7.

[12]Drug War Results 'Spotty,' *Washington Post*, April 14, 1990. p. A1.

Fig. 8.1. Top-of-the-line speculative housing in Great Falls, Maryland.

Social ecology

The explosive growth of metropolitan Washington has produced a good example of the 'galactic metropolis' described by Lewis (see Chapter 1 of this text). Seen from the air, much of the metropolis consists of hundreds of small subdivisions, business parks, commercial corridors and suburban nodes flung over rolling, wooded countryside and loosely strung together by tendril-like access roads and a web-like highway system. Within this physical template, the overall social ecology of the metropolitan area is best captured by the 'lifestyle communities' identified by consumer research. Figure 8.2a–c shows the spatial distribution of the area's major lifestyle communities identified by the Claritas Corporation.[13] Of the 40 lifestyle communities identified from national analysis, just nine account for nearly 80 percent of the Washington area's 1.4 million households.

[13] The company's national data are described and summarized by M. Weiss, 1988a. The following description of Washington-area clusters is based on his article 'What Your Address Says About you,' *The Washingtonian*, December 1988, pp. 158–61, 254–60.

The wealthiest communities, 'Blue Blood Estates,' include many of the neighborhoods listed in Table 8.1, including Potomac, Great Falls, McLean and Chevy Chase. These communities account for only six percent of the area's total households but they set the standards for its materialism. They are remarkably localized, straddling the Potomac in northern Fairfax County and southeastern Montgomery County.

The largest single group consists of communities of 'Young Influentials'—predominantly young, upwardly mobile singles and dual-career couples in white-collar jobs with substantial incomes. They live in places like Gaithersburg, Greenbelt and Parkfairfax, in neighborhoods strung along the major highways, with a high density of new and expensive townhouses and mid-rise apartments such as those along 'Condo Canyon' parallel to I-95 in Fairfax County. Marketing studies portray them as the exemplars of Ehrenreich's (1989) yuppies: they are more than twice as likely than average Americans to go sailing, take cruises, drink bottled water and attend concerts. Their favorite automobiles are BMWs, Acuras and Alfa Romeos. They spend twice as much time as the rest of the population exercising (jogging, sailing or playing racquetball or tennis are preferred) and have little time left for television. They subscribe in large numbers to *Gourmet* magazine but are so busy that they are the area's top customers for home-delivery pizza, Chinese take-out meals and upscale frozen dinners.

The second largest group of communities—'Furs and Station Wagons'—is dominated by upwardly mobile couples in their thirties and forties with teenage children. Their neighborhoods surround the Beltway to the northwest and southwest of the city and are characterized by recently-built subdivisions with amenities such as tennis courts, swimming pools and bike paths. Two-thirds of the residents of these communities have moved into their homes in the last five years, and three-quarters of them own three or more cars, 'the better to take their kids to piano lessons and soccer practice' (Weiss, 1988b, p. 254).

'Young Suburbia' communities, which account for 10 percent of the area's households, are dominated by younger households with elementary-school-aged children, on the first rung or two of the home ownership ladder, living in cheaper, high-density subdivisions in outlying suburbs like Clinton, Dale City, Herndon and La Plata. These households tend to drive inexpensive automobiles and shop in factory outlets and do-it-yourself stores but are attentive to future upward mobility, subscribing in large numbers to *Money* magazine.

Two groups of communities are dominated by suburbanizing black households. The more affluent 'Black Enterprise' communities such as Brookland, Capitol Heights and Walter Reed are scattered along the northeastern and southeastern boundaries of the District and in two suburban corridors in Prince George's County (Fig. 8.2c). These neighborhoods are middle class, with the majority of homes worth, in 1988, between \$80,000 and \$150,000. Like white middle-class suburbanites, their residents are distinctive in marketing terms for their membership of book clubs, their propensity to exercise, their investments in annuities and their purchasing of lawn furniture, TV sets,

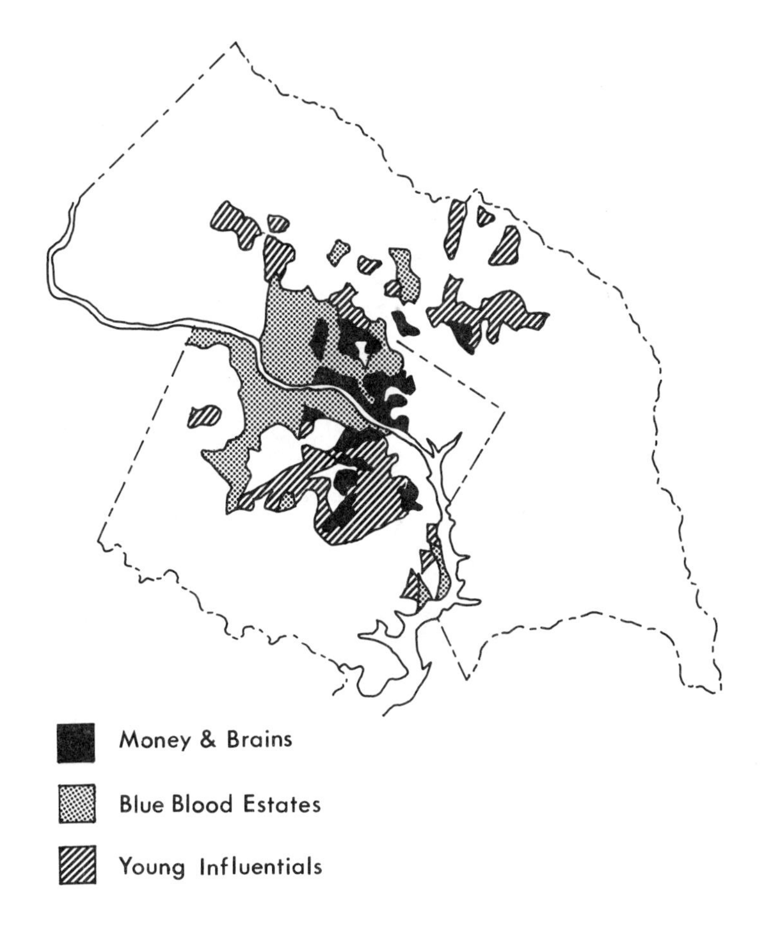

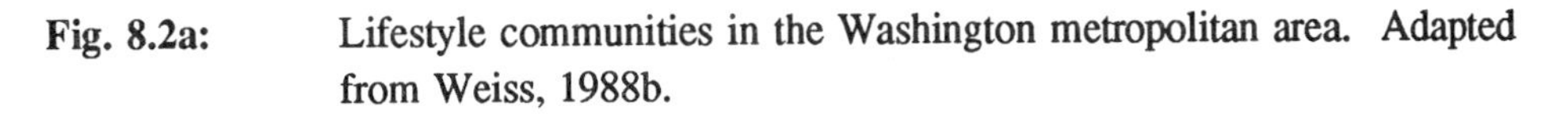
Fig. 8.2a: Lifestyle communities in the Washington metropolitan area. Adapted from Weiss, 1988b.

VCRs, compact disc audio systems, personal computers and movie cameras. Relatively less affluent are the 'Emergent Minorities,' communities of predominantly black working-class households located in the middle zones of the District and in outlying enclaves within the Beltway in Prince George's County. Examples include Brentwood,

Columbia Heights and Palmer Park. These are neighborhoods of rowhouses and apartment buildings where consumer tastes are characterized by above-average rates of malt liquor and menthol cigarette consumption, watching professional wrestling on TV and driving Renault Alliances, Chevy Novas, Yugos—and Cadillacs.

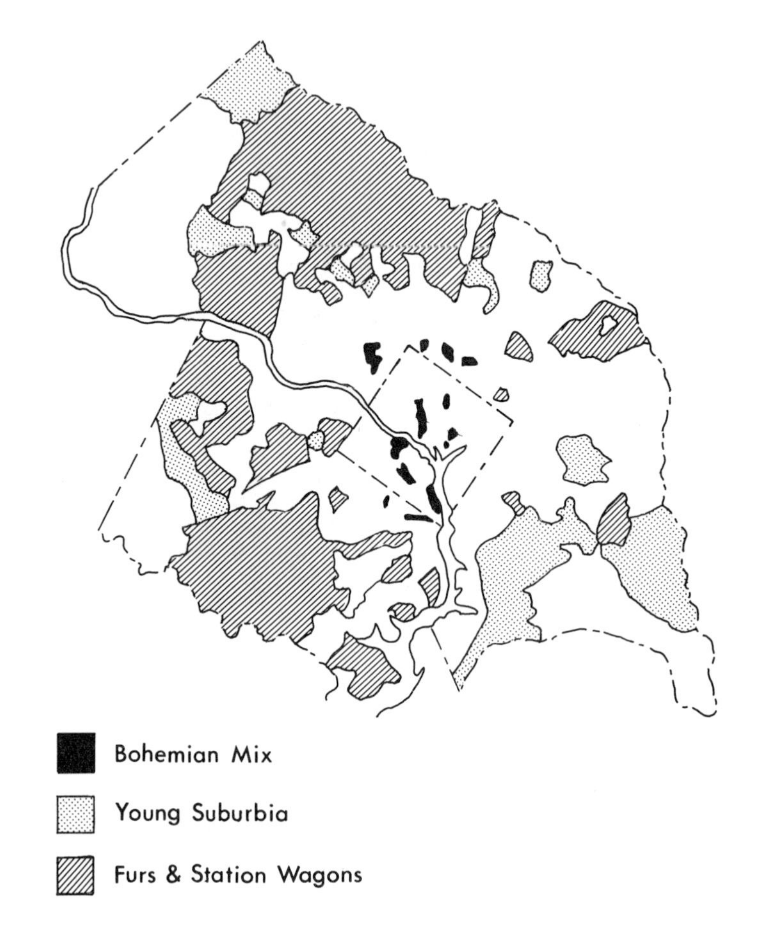

Fig. 8.2b. Lifestyle communities in the Washington metropolitan area. Adapted from Weiss, 1988b.

A third group of black communities consists of aging inner-city slum districts and housing projects—'Downtown Dixie Style.' These areas are characterized by unemployment rates of around 30 percent, low educational achievement and high levels of poverty, violent crime and drug abuse. They are located to the north and northeast of the CBD and in the southeastern parts of the District: typical examples include the area around 14th and U streets, Fort Chaplin Park, and the projects of Anacostia. In marketing terms, they are of significance only for their propensity for renting television sets, buying inexpensive subcompact automobiles, reading *Soap Opera Digest* and cashing promotional coupons for check-cashing businesses and high-risk insurance firms.

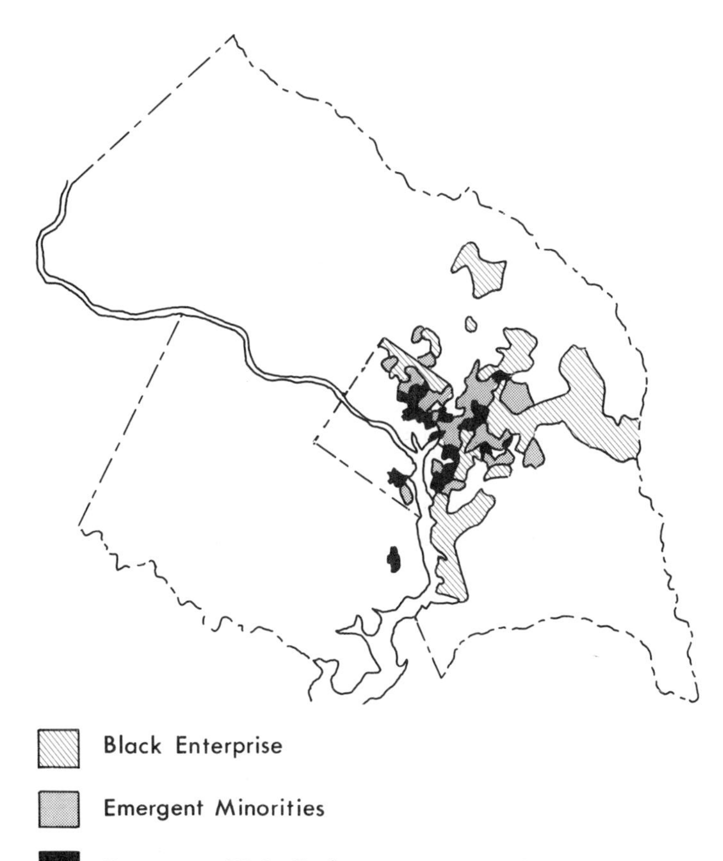

Fig. 8.2c: Lifestyle communities in the Washington metropolitan area. Adapted from Weiss, 1988b.

Finally, the Washington metropolitan area contains two groups of communities that encompass elements of gentrification. 'Bohemian Mix' neighborhoods, mostly within the District, contain pioneer, first-stage gentrification and transitional neighborhoods as well as more established but less affluent gentrified neighborhoods. Exemplified by Foggy Bottom and Dupont Circle, they are dominated by singles and divorcées, young and old, black and white, who are drawn to the urbanity of sidewalk cafes and storefront galleries set among brownstones and apartment houses. 'Money and Brains' neighborhoods, on the other hand, are characterized by high-income households living near urban university campuses. Some, like Georgetown, are dominated by upscale, gentrified row houses and their infill look-alikes; others, like Spring Valley, are dominated by older, established upper-middle-class housing. They contain large numbers of accountants doctors, lawyers and scientists—Bourdieu's 'new bourgeoisie' (see Chapter 1), who tend to buy European luxury automobiles, designer clothes and appliances, invest in stocks, bonds and securities and read *New Yorker* magazine.

The landscapes of postmodernity

Embedded within and superimposed upon this social ecology is a matrix of emergent landscape elements. Some, such as postmodern architecture and the packaged environments of mixed-use developments, appear throughout extensive sections of the metropolitan fabric. Others, such as gentrified neighborhoods and the refugee landscapes of isolated ghettoes, are restricted to particular localities. The remainder of this chapter describes the various material landscapes of postmodernity in more detail, paying attention also to the emergent spatiality and socio-spatial dialectics surrounding particular fragments in their ecological context.

Postmodern architecture

Postmodern architecture is eminently suited to the materialism and image-consciousness of Washington's new industries and their employees. In contrast to the abstract functionalism of modern architecture, postmodern buildings are scenographic, decorative and semiotic (Table 8.2). All this is attractive to developers with upscale businesses and consumers in mind: they know that although it costs more to build a 'rich' building, it will sell or rent more quickly—and often at a premium—because it can project the appropriate 'look.'

Yet in reading the landscape for its postmodern architectural content there is a fundamental problem. In the obtuse and fickle discourse of architectural 'theory' and criticism, postmodern architecture can be almost anything you want it to be, provided you want it badly enough. Postmodern architecture is by definition wide-ranging and eclectic, anarchical and combinatorial. As we saw in Chapter 1, one of its principal characteristics is 'double coding,' combining Modernist styling or materials with something else, usually historic or vernacular motifs. This allows the deployment of the iconography of everything from the neoclassical, Beaux Arts and Art Nouveau to the International Style, in which modern architecture itself is the object of irony and pastiche. Postmodern architectural style is the style of styles. It is this self-conscious stylishness that makes it attractive to the professional middle classes, that makes it such an important component of contemporary commodity aesthetics and that makes it so difficult—and dangerous—to pin down and catalog.

Within Washington, examples can be found of most of the major stylistic variants of postmodern architecture (Knox, 1991; Crow, 1990). The most common themes are what Jencks (1986) would describe as 'straight revivalism' (accurate reconstructions of period styles), 'postmodern classicism' (a freestyle classicism that goes beyond straight revivalism to re-work the vocabulary of classical architecture) and 'historicism' (allusions to period styles eclectically mixed with contemporary images and references). The former is particularly favored in the infillings and extensions of commercial space in the old downtown, such as the reconstruction of the Evening Star Building on Pennsylvania Avenue and the extensions to the Southern Building (between 15th and H Streets). The Southern Building is the work of a Washington architectural firm, Shalom Baranes Associates, that has specialized in expanding older office buildings by giving them added height in the style of the original structure. Postmodern classicism is also favored in downtown locations because of the opportunities it provides for deploying the motifs of status and the materials of opulence. Examples include 1201 New York Avenue (Fig. 8.3), 717 14th Street ('the renaissance,' according to its advertising, 'of a classic address'), and the Raleigh Building at Connecticut Avenue and DeSales Street NW—based on the formula of a Beaux Arts building (with a definable bottom, middle and top), with vertical fenestration and echoes of historical ornament (such as pilasters topped by blind balustrades and entrances set apart with stone pediments supported by archly elongated brackets) but with a blend of contemporary stylistic collage and curtain-wall glass in the middle parts of the façade. Most prevalent of all are historicist styles, such as in Republic Place at 1776 I Street (inspired by Louis Sullivan and the Chicago School, with an octagonal corner tower crowned—quite literally—by a loggia), the new Canadian Chancery (Fig. 8.4) on Pennsylvania Avenue (that echoes the nearby Municipal Center and the Jefferson Memorial) and One Cambridge Court, a 220,000-square-foot office building in Fairfax County (with a design that echoes early Gothic, incorporating a 95-foot glass-spired atrium with white, light- and dark-green marble and black granite in a mosaic patterned after one of the Medici palaces in Venice—specially researched by consultant architectural historians).

Table 8.2. Contrasting Characteristics of Modern and Postmodern Architecture

Modern	***Postmodern***
'Less is more' (Mies van der Rohe)	'Less is a bore' (Robert Venturi)
International style, or 'no style'	Double-coding of style
Utopian and idealist	Real world and populist
Abstract form	Responsive and recognizable form
Deterministic form	Semiotic form
Functional separation	Functional mixing
Simplicity	Complexity and decoration
Purist	Eclectic
Pro-technology	Disguised technology
No historical or vernacular references	Mixed historical and vernacular palette
Innovation	Recycling
No ornament	'Meaningful' ornament
Context ignored	Contextual cues
'Dumb box'	Scenographic

After Jencks (1977) and Punter (1988)

Together with a scattering of more radical postmodern designs (such as Washington Harbor (Fig. 8.5),[14] Tycon Towers (the 'Shopping Bag' building) and 2070 Chain Bridge Road (the 'Toilet Bowl' building)), such buildings represent the signature structures of postmodernity within the built environment. But postmodern architecture has penetrated much further into the fabric of the city than this. Commercial and residential townscapes throughout the metropolitan area are suffused with the watered-down tropes of postmodern design. New single-family homes are built in revival styles or historicist modes, while shoebox stores and offices are peppered with arches, atria, columns, pediments, keystones, pilasters, semicircular windows, cornices, balusters, urns and cupolas; and office villages and shopping centers have come to resemble period stage settings—'Lite' architecture, the built environment's equivalent of easy-listening music, lite beer and lean cuisine (Goldberger, 1990).

[14]Known in local circles variously as the Martian Embassy or as one of the Weird Sisters of the Potomac (the other two being the Watergate Center and the Kennedy Center).

Fig. 8.3. 1201 New York Avenue, Washington, D.C.: Postmodern classicism.

Such outcomes appal many of the architectural cognoscenti. Harris and Lipman, for example, write that 'Aesthetic form is being emptied of social content. Post-Modern maturity is upon us' (1988, p. 98), concluding that 'Architecture, the craft of making buildings, has been reduced to an alliance of taste and capital, of art and profit, of style and power' (p. 107). Such observations miss the point, narrowly but decisively. The social content is there, but it is not the utopian social content of modernism. It is the materialistic and hedonistic social content of the status- and consumption-oriented society described in Chapter 1. The craft of making buildings has not been *reduced* to an alliance of taste and capital but has merely been sustained by *new* alliances of taste and capital.

Fig. 8.4. The Canadian chancery in Washington: decentered and double-coded.

Packaged landscapes

These new alliances, and the materialism upon which they are based, are inscribed most clearly on the urban fabric in the packaged landscapes of mixed-use developments (MXDs) and private master-planned communities. These are the 'artful fragments' (Boyer, 1990) of postmodern urban development, serially-produced set pieces that reflect the logic of corporate consolidation, flexible strategies, product differentiation, and public-private cooperation (see Chapter 1). According to the Urban Land Institute's Real Estate Project Database, there were fewer than 25 MXDs in the entire United States at the end of the 1960s. Construction began on 65 more during the 1970s and on nearly 300 more during the 1980s (ULI, 1989). This proliferation is largely the result of the restructuring and repositioning of the development industry. The increased involvement of large corporations and financial institutions in real estate made large amounts of capital available for prestige projects, while developers were impelled to put together projects that could support the rising costs of urban land while spreading design and management costs over bigger units and the same time incorporating flexibility to differentiate space for different categories of users.

Fig. 8.5. Washington Harbor, a mixed-use development, facetiously referred to as 'The Martian Embassy.'

In the Washington metropolitan area there are more than 40 MXDs,[15] two-thirds of which were initiated in the 1980s. They have been fostered by public agencies who view

[15]Including multi-use developments.

them as attractive additions to the tax base, as anchors for urban planning and design projects and for revitalization programs, as useful transition elements between different land uses and as potential stimuli, when located near Metro stations, for ridership on the public transit system. Indeed, some MXDs, such as National Place, have been publicly initiated. The District of Columbia was one of the first central cities in the United States to relax single-purpose zoning when, in 1974, it introduced a mixed-use zoning ordinance for twelve blocks of the West End (to the north of Pennsylvania Avenue and to the east of Georgetown), providing incentives and bonuses for moderate-income housing, pedestrian or cycling areas and retail or service space that might contribute to the vitality of the area. More recently, the District has encouraged mixed-use development in the traditional retail-commercial sector of downtown[16] with a zoning overlay that requires 18 to 20 percent of a new building's gross first floor area to be preferred retail, service or arts uses. Meanwhile, the Pennsylvania Avenue Development Corporation has assembled over 110 acres of land along the city's most important symbolic axis (between the U.S. Capitol and the White House) and established a planning framework that includes a number of MXDs, including National Place, the Willard Hotel complex, the Old Post Office (Fig. 8.6) and Market Square. Altogether, eleven of the 40 MXDs in the Washington area have been assisted by significant changes in public codes or ordinances; nine have been assisted by direct public investment in supporting elements such as access improvements, garages, arenas, libraries and cultural facilities; and five have been assisted by direct public ownership or shared risk.[17]

The net result is a landscape studded with clusters of glitzy nodes. About one third are located in the CBD, with the rest in commercial nodes and corridors and in peripheral edge-city concentrations convenient to 'Blue Blood Estates,' 'Young Influentials' and 'Furs and Station Wagon' neighborhoods. The micro-landscapes of the MXDs themselves are carefully packaged, with striking architecture, luxurious-looking materials and fittings, and interior spaces planted like rain forests. Upmarket commercial and residential tenants and fashionable retail outlets are accompanied by health centers, small convenience stores, concierge services and security patrols. Some have exclusive restaurants and business clubs, fitted out in mahogany, crystal and lush carpeting. A typical example is the Twinbrook Office Center/Holiday Inn Crowne Plaza in Rockville, Maryland. Located on a 6.6-acre site across from the Twinbrook Metro Station, this $55 million, 700,000-square-foot project includes 12,400 square feet of ground-level retailing, a 315-bed hotel, two restaurants, a 19,000-square-foot conference facility, an indoor/outdoor pool and a

[16]Bounded on the west by 15th Street, on the north by New York Avenue and H Street, on the east by 9th Street and on the south by E Street and Pennsylvania Avenue

[17]Data from the Urban Land Institute Project Data Base, 1990.

Fig. 8.6. The Old Post Office, now converted to a mixed-use 'festival' space.

racquetball and health club. The architecture combines a Late Modern exterior with a lush, postmodern interior space that features an atrium with a large wooden Victorian-style pavilion set amid waterfalls, ferns and meandering brick walkways. Settings such as this are an important component of the postmodern urban matrix because of their visibility and exclusivity. Like larger and even more spectacular settings based on waterfront redevelopments or festival marketplaces, they are part of the 'carnival mask' of contemporary urbanization (Harvey, 1989, p. 92), settings that serve as a focus of stylish materialism while diverting attention from the refugee landscapes of the poor and the homeless.

A second important form of packaged landscape is represented by the private master-planned communities that have proliferated within the 'Blue Blood Estates' and 'Furs and Station Wagon' ecology of Washington's northern and western suburbs.[18] They are a response to the mass production and mass consumption of split-level suburban space, a result of product differentiation and carefully-targeted niche marketing. By exploiting PUD zoning (see Chapter 1), developers can put together projects that are attractive to a very profitable sector of the residential market while retaining scope for flexibility in the composition and timing of the development. Residents of such communities are offered sequestered settings with an extensive package of amenities that typically includes tennis courts, a golf course, swimming pools, play areas, jogging courses, an auditorium, exercise rooms, a shopping center, a day-care center and a security system symbolized by imposing gateways and operated by electronic card-key systems. Housing is typically a mixture of expensive single-family homes, upscale townhouses and condominiums and a few smaller studios or apartments for young singles or the elderly, all in High Suburban style: mock-Tudor, mock-Georgian, neo-Colonial, Giant Cape Cod and so on. The entire ensemble is framed in a carefully landscaped setting that might contain a lake stocked with swans or a neoconservationist assemblage of remnant woodland, an artificial wetlands environment and plantings of wild flowers. The landscape is completed by a parade of joggers in expensive warm-up suits and by busy UPS vans delivering affordable luxuries from the mail-order branches of Spiegel, Williams-Sonoma, Sharper Image and Bullock and Jones.

These artful fragments provide the ultimate framework for the culture of stylish materialism. They are communities of affect rather than communities of interest, sharing only a dedication to the iconolatry of visible wealth and distinction, a sect that has adopted the architecture of the English gentry, the artifacts of French aristocracy and imperial China and contemporary European kitchen and automobile technology as signs of membership. The very names of the communities are carefully selected in order to set the required tone of distinction, heritage and authenticity: Hampton Chase, Lansdowne, Manor Gate, River Oaks, Stilloaks, Tavistock, Woodlea Manor and so on. Advertising imagery draws on the totemism of golf, equestrianism and pastoral landscapes, while

[18]The essential features of these communities are: 'a definable boundary; a consistent, but not necessarily uniform, character; overall control during the development process by a single development entity; private ownership of recreational amenities; and enforcement of covenants, conditions and restrictions by a master community association' (Suchman, 1990, p. 35).

advertising copy leaves no doubts about the status and stylishness of the product. Woodlea Manor, for example, is described as 'Exclusive. But not entirely out of reach,' and as a place 'Where style of living matters.' The slogan for Sully Station is 'Sometimes It's Better to Live in the Past;' for King's Forest ('The Community That Says "You've Arrived" Has Arrived') it is 'Your Crowning Achievement.' Seven Lakes, a planned community created by the Devon Land Corporation, is advertised as follows:[19]

> *Imagine owning a home of unbridled quality crafted by one of Northern Virginia's finest builders. A magnificent home destined to stand on almost four acres of unspoiled countryside in Stafford County, Virginia. Imagine owning your own paddock. Even a barn. Or using the complete facilities of a beautifully appointed equestrian center with a dressage ring and a jumping and turn out area.*
> *Just imagine fishing on Wordsworth Lake. Relaxing in a stone built clubhouse with lounges, a bar, even a sauna. Or working out in a fully equipped exercise room. And just imagine your surprise to find an Olympic-size swimming pool—surrounded by trees—a stone's throw away.*
> *Imagine it all. With seven well stocked crystal clear bodies of water. Seven Lakes. If you want this life. Come see it now.*

Within the Washington metropolitan area there were in 1989 some 30 private master-planned communities (including Reston, a suburban new town in Fairfax County built by R.E. Simon in the 1960s and subsequently taken over by Gulf Oil and then Mobil Oil; and Columbia, a suburban new town in Howard County built at the same time by J. Rouse, developer of Baltimore's Harbor Place, both now enjoying an expansion with the popularity of packaged residential settings) (Knox, 1991). Most of them were begun during the development boom of the late 1980s, and many are still incomplete, waiting out the recession of the early 1990s.

Reclaimed landscapes

In many ways, the reclamation of older fragments of central-city areas through preservation and conservation can be attributed to the failure of urban redevelopment schemes. Citizen protests against wholesale demolition and renewal were an important part

[19]From a full-page advertisement in the *Washington Post*, September 23, 1989, p. E31.

of the broad countercultural movement of the 1960s. Their success helped to redefine the conventional wisdom among city planners and urban policymakers. By the 1980s, historic preservation had become part of an officially-sponsored 'heritage industry,' bolstered by tax credits and accelerated depreciation benefits for investments in historic property (Gleye, 1988). Between 1970 and 1985 the number of properties and districts listed in the National Register of Historic Places increased from 1,500 to 37,000 (Listokin, 1989).

Yet it is no coincidence that historic preservation has been a striking aspect of urban change during the blossoming of postmodernity. Historic buildings and districts lend distinctiveness and identity to both residential and commercial users while resonating very clearly with those aspects of postmodern culture that emphasize the past, the vernacular and the decorative. Furthermore, the heritage industry is bound up in the struggle for cultural legitimacy and at the same time is an important component of the 'society of the spectacle.' Historic preservation not only reclaims buildings and neighborhoods from the ravages of physical decay but, much more significantly, reclaims them from the ignominy of social decline. Occupants of refurbished buildings are able to draw on the cultural capital and social prestige of earlier occupants; meanwhile, the buildings themselves become part of the scenography of the contemporary city.

In Washington, urban renewal programs had been particularly vigorous during the 1960s and early 1970s. President Kennedy had expressed his dismay at the urban decay visible along the route to his inauguration, and the Pennsylvania Avenue Development Corporation was subsequently established in order to tidy up and revitalize the axis. Then, after the riots and civil disorder of the late 1960s, office and retail activity were repelled from the old downtown area to the east of the White House, prompting the redevelopment of a 'new downtown' area to the north and west (Gale, 1987). In the process, the demise of several familiar landmarks prompted the formation of an activist preservation group, Don't Tear It Down, which soon became an important element in local politics. The first victory of the group (now known as the DC Preservation League) was the prevention of the demolition of the Old Post Office, which was subsequently restored and adapted to a mixed-use 'festival' space that is a key element in the Pennsylvania Avenue Development Plan. Other notable victories for the group have included the preservation of the Willard Hotel and the nineteenth century terrace known as Red Lion Row, both on Pennsylvania Avenue.

By the time the old downtown area began to be ripe for reinvestment in the mid-1970s, Washington's preservationist movement had become a considerable force, and developers had come to recognize the demand for historic settings. As a result, the downtown townscape is now a virtual set piece of preservation projects, including Gallery Row, the National Bank of Washington, the Hotel Washington, the Evening Star Building, Woodward and Lothrop's, the Colorado Building (Fig. 8.7), the Sun Building, the McLachlen Bank, the Southern Building, the National Theater and the Warner Theater

(Knox, 1991). Overall, almost 4000 structures have now been designated under local preservation laws, twelve areas have been designated as residential historic districts, and an entire volume of the District's Comprehensive Plan is devoted to historic preservation (Gale, 1987).

Fig. 8.7. The Colorado Building: renovated office space in the old downtown.

The success of historic preservation fostered the spread of gentrification—a kind of do-it-yourself postmodern design that has enabled undervalued central city neighborhoods to be 'reclaimed' by capital through the invasion of young professionals seeking to establish a distinctive lifestyle and habitus (Jager, 1986; Mills, 1988; Smith, 1987; Zukin, 1988). While historic preservation has had little direct effect on the value of adjacent land or buildings (Gale, 1989), designated buildings and districts have been important components of the avant garde *mise-en-scène*, part of a diorama assembled from 'real' places with 'real' buildings and 'real' people. This 'reality' is, of course, one of the first victims of the socio-spatial dialectic that accompanies gentrification. 'Real' people are displaced, 'real' buildings are remodelled, and 'real' places are transmogrified into isolated, sanitized enclaves of urbane materialism.

In Washington, these enclaves include parts of Georgetown and Old Town Alexandria (both the object of gentrification for more than 50 years), together with a broad crescent of neighborhoods that encircle the CBD to the north. They encompass a broad spectrum of landscapes that vary according to the stage of gentrification. In the mature gentrified 'Money and Brains' neighborhoods of Georgetown, Old Town Alexandria, the Kalorama Triangle area and Capitol Hill, Georgian and Federal residences valued at $300,000 to $600,000 are interspersed with boutiques, flower shops and antique stores. The neighborhoods are introverted: there is little street life, and the character of the area has to be read from the telltale glimpses of interior décor, the correctness of exterior paint and trim and the dominance of expensive European automobiles parked on the street.

At the other extreme, in the vanguard of gentrification in 'Bohemian Mix' neighborhoods such as Shaw, Mount Vernon Square and Dupont Circle, it is the reclaimed homes (early twentieth-century rowhouses and 3- or 4-story walk-up apartments) that are interspersed—among dilapidated homes, crack houses, X-rated businesses and second-hand stores. These neighborhoods are extroverted, their vibrancy an important attraction for pioneer gentrifiers. The landscapes of these first-stage gentrifying areas are captured vividly by Michael Dolan (1990, pp. 21–22) in his description of Dupont Circle:

> *Dupont Circle is where, when the geeky bespectacled kid on the skateboard isn't about to flatten your arches, the mumbly bum is begging you for half your burrito. It's where the quiet comes and goes amid drunks blustering themselves into another dust-up and the Park Service groundsman revving the motor on his rototiller. It's where the skinheads stride by with their boomboxes and their big black boots as primly grim success-dressers ignore the lean tattoo-faced fellow haranguing them for spare change when he isn't participating in mysterious triangular transactions involving tightly folded currency and small, possibly contraband packages . . . .*

> *It's the place where neighborhood denizens and passers-through . . . shop the daily specials at Larimer's grocery, following the European tradition of buying provender a meal at a time. It's the place where you can pick up a balalaika for $69.99—a "glasnost special"—at Ardis Music Center, or climb four flights upstairs and study tai ch'i at Great River Taoist Center, then stroll a few doors north to Food for Thought, the continually endangered eatery, on whose 20-foot bulletin board "Returned Peace Corps Volunteer (Nepal) & George Washington Graduate Student (Non-Smoker, Liberal) Seeks Housing" coexists with "Two People Wanted for Rooms in a Negative Force Steeljaw Trap House With Six Hunks and One Hussy 19-24 (We Like Bourbon, Johnny Cash, and Fungus)" . . . .*

Refugee landscapes

Amid the residual enclaves of 'Downtown Dixie Style' neighborhoods and the more extensive tracts of 'Emergent Minority' neighborhoods in the District are landscapes where drug alleys, crack houses and shooting galleries are embedded among dreary housing projects and apartment blocks, the only refuge for the worst casualties of metropolitan economic and social transformation. While ghettos and slums are by no means new components of urban structure, these landscapes of the excluded are unprecedented in the intensity of combined poverty, violence, despair and isolation. Places that are avoided by everyone (even their own inhabitants), they are characterized by red brick duplexes and triplexes that bear the stigmata of boarded-up windows, bullet-holed doors, yellow police tape and spray-painted graffiti. Lawns and open spaces are strewn with debris; the only people who seem to spend any time out of doors are drug dealers, terminal-stage addicts and thick-legged prostitutes.

Potomac Gardens (Fig. 8.8), 12 blocks from the U.S. Capitol, is a typical example of such refugee landscapes. As described in a *Washington Post* investigation,[20] nearly one third of the 306 households in the complex in 1989 had to rely on public assistance. Only ten families contained both a father and a mother; two were headed by fathers, the rest by single mothers. There were 46 vacant apartments and about twelve apartments within the complex where crack was regularly bought and used. While drug use/non-use creates a critical social cleavage in such settings, the *type* of drug preferred by users provides the basis for finer social differentiations. '[Heroin] junkies despise the

[20] L. Duke and D. M. Price, A Microcosm of Despair in DC, *Washington Post*, April 2, 1989, pp. A1, A27–8.

"pipeheads" whose entire daily existence is often spent getting high or looking for rocks. Crack smokers, meanwhile, consider themselves superior to heroin junkies, who litter the complex with needles'[21]—and will shoot up anywhere, cooking their dope with rainwater from the ground and using the mirrors of parked cars to locate veins in necks or groins.

Fig. 8.8. Potomac Gardens, Washington, D.C. *Washington Post* photograph. Used with permission.

The going rate in 1989 for fifth graders paid by dealers to keep watch was $3; and a 12-year-old was widely known as one of the most aggressive drug dealers. 'A police source said officers have seen the boy with wads of cash and residents said the boy carries a 9-mm pistol in a shoulder holster.'[22] The link between drugs and violence is intimate. Potomac Gardens is part of an eruption of murder and drug dealing that covers much of the eastern half of the District (Fig. 8.9). Law-abiding residents of these areas, if not literally caught in the cross fire, are caught up in a desperately bleak socio-spatial

[21]*Ibid.*, p. A27.

[22]*Ibid*, p. A27.

dialectic in which both the built environment and its inhabitants are increasingly isolated and stigmatized. The everyday lifeworld of many residents has already come to accommodate having to watch cars carefully in order to avoid being caught in a drive-by shooting, to lock themselves into their apartments early in the afternoon, to draw shades and blinds, to watch television from positions well away from windows and to stuff towels around doors to keep out crack fumes from the stairwells.

Fig. 8.9. Drug markets and Homicides in the District, 1988. Redrawn from *The Washington Post*, 1/13/1989, p. E1.

Equally desperate are the landscapes and lifeworlds of the city's literal refugees—the homeless. Estimates of the number of homeless persons in the Washington metropolitan area range from 12,000 to 27,000. Most of the homeless are located in the District, although Fairfax County contains several thousand homeless persons[23] and other suburban counties each contain at least several hundred. Surveys suggest that single men account for only about 35 percent of the homeless; children account for more than half. The 'landscapes of despair' (Dear and Wolch, 1987) generated and inhabited by the homeless are particularly visible in the District, where the bundles and carts of homeless persons huddled over Metro grates and under freeways in winter and sprawled over the lawns and benches of public open spaces in summer make for a stark contrast with the architecture of the federal area. Despite the efforts of groups such as the Community for Creative Non-Violence (led by the late Mitch Snyder), the city's record in caring for the homeless is poor. It did have a policy, between 1984 and 1990, of providing unlimited overnight shelter to the homeless, but the shelters themselves have been shown to be characterized by filthy blankets, rats, lice and scabies, broken toilets and streams of running sewage (Gellman, 1989). Vulnerability, hunger and illness are the hallmarks of the lifeworld of the homeless; ephemerality, crowding and squalor are the hallmarks of the micro-landscapes they inhabit. Together, their lifeworlds and landscapes have become emblematic of the dualism of contemporary cities. Their continued existence, meanwhile, attests to the vicious exclusion that is the corollary of the hedonistic materialism of the postmodern city.

The rhetoric of postmodern landscapes

This last point brings us to the important question of the meaning of new landscape elements and their role in legitimizing and reproducing the social order. Duncan (1990) refers to the 'rhetoric' of landscapes in discussing the mechanisms by which signification takes place within the built environment. He suggests that 'By becoming part of the everyday, the taken-for-granted, the objective and the natural, the landscape masks the artifice and ideological nature of its form and content. Its history as a social construction is unexamined. It is, therefore, as unwittingly read as it is unwittingly written' (p. 19). Change in urban morphology occurs relatively slowly, particularly within the contemporary context of rapid technological change and time-space compression. The emergence of packaged, restored and refugee landscapes has therefore been almost completely unnoticed, while their significance in forming the basis of a new socio-spatial dialectic has been virtually unexplored.

[23] These include many working poor who cannot afford the rents in the area and are forced to camp in local parks such as Burke Lake Park and Lake Fairfax Park.

References

Boyer, C. 1990 The return of the aesthetic to city planning, pp. 93–112 in D. Crow (ed.), *Philosophical Streets*. Washington, DC: Maisonnueve Press.

Crow, D. 1990 *Philosophical Streets*. Washington, DC: Maisonnueve Press.

Dear, M. and Wolch, J. 1987 *Landscapes of Despair*. Princeton: Princeton University Press.

Dolan, M. 1990 A short history of a very round place, *Washington Post* Magazine, September 2, 1990, 18–39.

Duncan J. A. 1990 *The City as Text*. Cambridge: Cambridge University Press.

Ehrenreich, B. 1989 *Fear of Falling*. New York: Pantheon.

Gale, D. 1987 *Washington, DC: Inner City Revitalization and Minority Suburbanization*. Philadelphia: Temple University Press.

Gale, D. 1989 The Impact of Historic District Designation in Washington, DC. Occasional Paper #6, Department of Urban & Regional Planning, George Washington University, Washington, DC.

Garreau, J. 1991 *Edge Cities*. New York: Doubleday.

Gellman, B. 1989 D.C. Ordered to Speed Changes on Shelters, *Washington Post*, January 8, 1989, A1–A19.

Gleye P. H. 1988 With heritage so fragile: A critique of the tax credit program for historic building rehabilitation, *Journal of the American Planning Association*, 54, 428–88.

Goldberger, P. 1990 After Opulence, a new 'Lite' Architecture, *New York Times*, May 20, 1990, Section 2, pp. 1, 38.

Granat, D. and Conlin, J. 1990 The Sky is Falling? *Washingtonian*, 25, 7, 88–96.

Harris, H. and Lipman, A. 1988 Form and content in contemporary architecture—issues of style and power, *Architecture et Comportement*, 4, 97–110.

Harvey, D. W. 1989 *The Condition of Postmodernity*, Oxford: Blackwell.

Jager, M. 1989 Class definition and the aesthetics of gentrification: Victoriana in Melbourne, pp. 78–91 in N. Smith and P. Williams (eds.), *Gentrification of the City*. Boston: Allen and Unwin.

Jencks, C. 1977 *The Language of Post-modern Architecture*. New York: Rizzoli.

Jencks, C. 1986 *What is Postmodernism?* New York: St Martin's Press.

Knox, P. L. 1987 The social production of the built environment: Architects, architecture and the post-Modern city, *Progress in Human Geography*, 11, 354–78.

Knox, P. L. 1991 The Restless Urban Landscape: Economic and Socio-Cultural Change and the Transformation of Washington, D.C., Annals, *Association of American Geographers*, 81, 181–209.

Listokin D. 1989 Landmark designation: An emerging form of land use control, Paper presented to annual meeting of the Association of Collegiate Schools of Planning, Portland, Oregon.

Mills, C. A. 1988 Life on the upslope: The postmodern landscape of gentrification, *Society and Space*, 6, 169–89.

Parham, L. 1990 Washington, DC, *National Real Estate Investor*, 32, 6, W1–W16.

Punter, J. 1988 Post-Modernism: A definition, *Planning Practice and Research*, 4, 22–8.

Smith, N. 1987 Of yuppies and housing: gentrification, social restructuring, and the urban dream, *Society and Space*, 5, 151–72.

Suchman, D. R. 1990 Housing and community development. In D. Schwanke (ed.), *Development Trends 1989*. Washington, D.C.: Urban Land Institute, 28–37.

Urban Land Institute 1989 *Mixed-Use Development Handbook*. Washington, D.C.: Urban Land Institute.

Van Dyne, L. 1989 Money fever, *The Washingtonian*, October 141–5, 239–56.

Waldrop, J. 1990 Shades of black, *American Demographics*, 129, 30–33.

Weiss, M. 1988a *The Clustering of America*. New York: Harper and Row.

Weiss, M. 1988b What your address says about you, *Washingtonian*, December, 158–61, 254–60.

Weissman, E. 1990 The facts of life, *Regardie's*, February, 164–74.

Zukin, S. 1988 *Loft Living, Culture and Capital in Urban Change*. New York: Radius Books.

Nine

BERLIN'S SECOND MODERNITY

Scott Lash

Lancaster University

Genesis

Georg Simmel and Walter Benjamin were more than just the first social interpreters of modernity; more than just the original allegorists of a newly restless urban landscape; more than just excluded creative figures on the margins of early twentieth-century European thought. Simmel and Benjamin were *Berliners*. And when Walter Benjamin and Georg Simmel wrote about the Berlin of their youth, what they were talking about clearly was a *Weltstadt*.[1] Berlin was at the end of World War I, when it had opened out its borders for probably the last time to incorporate former suburbs, already a city of 4 million people—and, as such, by far the fastest growing city in Europe. From 1871 to 1913 today's most significant sections of Berlin—Charlottenburg, Schöneberg, Wedding in the West and Prenzlauerberg and Friedrichshain in the East—increased on the average some seven- to tenfold in population. Even the old proletarian districts right in the center of the old Berlin—the former GDR's East Berlin—increased often two- to fourfold in population during this time.

Berlin was a culture center and was becoming a *modern* culture center. It was already—half a century before the Wall—a city of two centers. Alongside the old center ranged in the vicinity of Potsdamer Platz and Leipziger Platz, in which business was booming but in which the state played an important architectural, cultural and economic role, grew Berlin's second center along Tauentzienstrasse and Kurfürstendamm. This second center did not have the state buildings. It was predominantly privately owned and regulated, including its theater, its art galleries and the like. And it was very modern, very avant-garde. It sprouted, for example, alongside the old university (now the Humboldt University) in the old center of East Berlin, two new universities in the West,

[1] *Weltstadt*: World city, global city.

the Technical University just on the northwest borders of the Kurfürstendamm *Innenstadt*[2] and the Freie Universitat well to the southwest in Dahlem (Lash, 1990).

The Berlin of Simmel and of Benjamin was Germany's unparalleled economic center. Meso-economic shopping districts spanned both centers on the eve of World War I. Emblematic here was the imposing Wertheim department store in the Leipziger Strasse in the East, built by the great Alfred Messel. And the KDW (Kaufhaus des Westens) near Wittenbergplatz in the West, the symbol today of West German, indeed Western, consumer society for 1990s crowds of 'Ossis' (East Germans) and Poles. But Berlin was also Germany's unparalleled industrial leader—the modern electronics sector was located here, both Siemens and AEG. It was Germany's banking center, it's newspaper and printing and magazine center. Clothing industries and engineering also thrived.

Simmel's and Benjamin's Berlin was the West's, if not the world's, fastest-*modernizing* city. To take a cue from Alexander Gerschenkron (1962), Berlin exemplified *urban* 'backwardness in perspective'—in perspective, because it could draw on the modernizing experiences of a number of other already modern western metropolises. Berlin's urban development experience resembled the industrial experience of latecomers, who could begin industrialization in a far more modern state than the early developing British and French. Berlin did indeed have a pre-modern core. But this was tiny in comparison to that of Paris, London and numerous Italian cities and more comparable to the pre-modern spatial density of cities in the New World such as New York or Boston. Starting almost from scratch, the whole thing could be incredibly modern all of a sudden and all at once.

Berlin was, as Sigmund Kracauer commented (Frisby, 1985), thus modern in not only being 'cleaned up' but also in being 'cleaned out.' And modernization as 'cleaning out' is an important component of the process of 'time-space distanciation,' about which Anthony Giddens has written so insightfully. That is, that space becomes, in modernization, more homogenous, more mediated, more universal, as the particularistic markers of the village and the pre-modern city are 'emptied out' in the contemporary urban landscape (Giddens, 1984, 1990). Paris, as Baudelaire commented, with its especial mixture of the ancient and particular and the modern of the boulevards is a far less emptied out or 'cleaned out' exemplification of urban space than Berlin. The military- and then the automobile-designed extraordinarily broad thoroughfares of Berlin—connecting the polycentric city localities—are actually closer to the 'emptied outness' of a sort of Los Angelesation (New York's 'full' avenues are far closer to the model of the Parisian boulevard) *avant la lettre*. These, at least at first glance, seem to be boulevards for traffic only. Flaneurs verboten!

[2] *Innenstadt*: City center, downtown.

The Berlin of Simmel and Benjamin was time-space distanciating at a more breakneck pace than elsewhere also in the sense of bringing the previously spatio-temporally disparate into contiguity. For example, the bringing together of previously disparate high culture and economic modernization was first accomplished in the development of breaking with the ancient curriculum for the modern and the innovation and institutionalization of the eminently modern Ph.D. degree at Berlin's Technical and Free Universities. Spatio-temporal distanciation was also exemplified in the incredibly important role of the previously distant Jews in Berlin's cultural life. Again these are advantages of urban (and cultural) backwardness in perspective. Cultural life and the universities in France and Britain were so 'full' of tradition that the supersession of the ancient by the modern curriculum has still, in important respects, not occurred. Paris and London as cultural centers were already so 'full' of Frenchmen and Britons that the spatio-temporally distant Jews could not hope to play the same sort of role as in the relatively 'emptied out' metropolises of New York, Berlin and, later, Los Angeles.

Exodus

The Berlin, then, that Simmel and Benjamin described, and perhaps, more importantly, *exemplified*, was indeed the most rapidly modernizing, the most hectically time-space distanciating, the most *restless* urban landscape of their time. Indeed if Paris was the capital of the nineteenth century, Berlin was the capital of the first decades of this century. But this process went into reverse gear from 1933, a process which was only accelerated from 1945. What happened effectively was that Berlin's already—in the best sense—emptied-out *symbolic* space was succeeded by the literal—and in the worst sense—emptying out of its *material* space. In the economy everybody left Berlin. Finance and the stock market went to Frankfurt, the print media went to Hamburg, and corporate headquarters relocated in Düsseldorf. Berlin had previously been the unparalleled world *Hauptstadt*[3] of the then new electro-technical sector. But now even electronics giants Siemens and AEG located their most advanced production establishments in West Germany, only the rump staying in Berlin. Further, Munich has become the center for high tech and Stuttgart, increasingly the engineering center.

Politically, too, Berlin underwent a literal and material process of 'emptying out.' The national capital, as Germany's political weight drastically contracted, went to Bonn.

[3] *Hauptstadt*: Capital.

The republic federalized. Berlin was under the aegis of the four *Besätzungsmachte*.[4] Berlin itself was of questionable political status in the Federal Republic, which subsidized it to a greater extent than America subsidizes Israel. Further, the political landscape was de-peopled as parties and trade unions, themselves de-politicized in comparison to Weimar Republic days, found their centers outside of the old *Hauptstadt*.

And this process took place culturally as well. Berlin's televisual production apparatus, for example, is rather insignificant in comparison to Nord-, Süd- and Westdeutschen Rundfunk and their locations in Hamburg/Hannover, Cologne and Stuttgart/Munich. And Berlin's literal emptying out, which was more a symbolic 'exodus' than a symbolic emptying out, also robbed the city of its cultural diversity. This is exemplified at its best in Berlin's postwar built environment, with its huge patches of empty wasteland—captured brilliantly in Wim Wenders' *Himmel über Berlin* (*Wings of Desire*) and in its featureless 1960s and early 1970s modernist high-rise buildings. Three political forces, acting consecutively, contributed to this wave of symbolic destruction—the Third Reich, the bombardments of the Second World War and the thoughtless city planners and architects of the initial postwar decades. This symbolic and material destruction—and time-space *compression* (in the sense of the negation of diversity)—was also finally exemplified in the forced exodus—one way or another—of the Jews.

As big an obstacle to time-space distanciation was the erection of the Wall and iron curtain. Berlin was constituted then as a time-space compressed 'island' forcibly separated from its hinterland, in the former DDR.[5] Thus East Berliners and other East Germans were spatially excluded from the Western *Stadtteil*,[6] and, moreover, the normal influx and ethnic mix of Poles, Russians, Czechs and others was cut off. Modernization is also, in the terms of classical sociological theory, a process of structural differentiation. And Berlin's hinterland was dominated by the anti-modernist hegemony of the Eastern Bloc, whose *Gleichschaltung*[7] precluded not just cultural realms, but also the economy from differentiating out of the political. With cultural modernization possible in the media, in architecture and in regard to ethnic diversity, Berlin's status as the world's cultural capital passed over to New York. Worse, Berlin, split into two and drastically time-space compressed, was no longer even a Weltstadt.

But the process of emptying out, the process of time-space compression into 'Island Berlin,' was an uneven one. It took place politically, economically and demographically

[4] *Besätzungsmachte*: Occupying powers, the Allies in Berlin after World War II.

[5] Deutsche Democratische Republik: the former East Germany.

[6] *Stadtteil*: Section of the city; urban district.

[7] *Gleichschaltung*: The drastic all-at-once executive order enforcing full centralization decreed by Hitler, eliminating all opposition and concentrating power at the top.

on the most massive of scales. Berlin was chopped into two; it was physically cut off from its immediate hinterland; it depopulated; its population aged. But in the cultural sphere this drastic process of spatial compression was not so clear at all. (West) Berlin became a sort of cultural mecca for West German political and aesthetic avant-gardes in the postwar period. This was first of all visible in the enormous physical presence of *students* in the city—and these are 'students' in a sense foreign to Anglo-American society. They begin their studies often at age 21 or 22 and take typically six to seven years to complete their degrees. They mix study with work, cafes, *Kneipen* and just plain hanging out. If we add in the significant number of *Doktoranden* (i.e., postgraduates, who do not in Germany come under the appellation *student*) and another group on the margins of academic life now often around forty years of age, who are completing their *Habilitation*,[8] you have a significant section of the population. The only comparable large city in the Anglo-Saxon world where students have such significant weight is Boston. The difference is that students do not dominate the central areas of Boston (partly because so many are outside in Cambridge) as they do Kreuzberg, Schöneberg and Charlottenburg in Berlin. That Boston has not the size nor the Weltstadt pretensions of Berlin. That Boston's students are kids set in well-organized frameworks of study, looking forward to a lucrative economic future, whereas in Berlin they are adults, only half-rooted in a chaotic university system and on a very pervasive margin of society. As urban planner and architect Wulf Eichstadt (1990) notes, many young people come to Berlin to study who do *not* want a future as a professional or manager living with two kids and wife at home in a Frankfurt or Hamburg suburb.

As an academic or an intellectual one can spend literally years in Berlin without having close contact with any real Berliners. The whole intellectual scene is *West* German. Your taxi driver will be an *echter* Berliner. But very few of those populating the cafes and Kneipen in Savigny Platz, Winterfeldplatz or in the Oranienstrasse will be. They will be almost exclusively from western Germany. And this is a very self-selective group. Among the men, many came because they did not have to do military service in Berlin. Others came because of radical politics, an aesthetic bent or particular sexual preferences. Berlin's Schöneberg has the perhaps apocryphal reputation of boasting the largest population of homosexuals outside of San Francisco. Berlin is above all not your *Yuppiestadt*. Anyone wanting suburban house, wife at home and two kids was to stay in Hamburg, Frankfurt, Stuttgart or Munich. Those rejecting the bourgeois virtues and Yuppiedom came to Berlin.

Further, Berlin boasted the theater, galleries, designer shops, classical, jazz and pop

[8] *Habilitation*: Further higher degree awarded in German-speaking world after the Ph.D. It requires another full dissertation and commonly eight to ten years work and is a requirement for a tenured professorship.

music scenes, the cinema selection of a world city. From Savigny Platz, just north and west of the Kurfürstendamm center of the Western *Innenstadtgebiet*[9] and within walking distance, say three-quarters of a mile, are probably 40 cinemas, 20 good restaurants, 15 interesting Kneipen, a large number of cafes, 30 designer shops, 25 art galleries and several first-rate theaters. And a lot of this culture is heavily subsidized. For example, in 1989, Robert Wilson's *The Forest* was produced at the Schaubühne in Lehniner Platz. With words by Heiner Müller and music by David Byrne, this play was performed on all four walls of the theater as well as in the air above the audience to a height of some 100–150 feet. And action in all five of these directions often proceeded simultaneously. This product unparalled in elaborateness played (to packed houses) only for three days. Estimated cost per spectator was 1000 marks ($670); actual cost per ticket was some 33 marks ($22).

If American culture was marked and formed by the generation of 1968 and English culture by the those experiencing their formative years a decade later in the ambience of punk, Berlin cultural life consists of large measures both of '*Acht-und-sechziger*' and punks. Sixty-eighters, with academic qualifications but unable—sometimes because of their politics—to find permanent teaching positions in the universities, came to Berlin and set up a host of independent intellectual magazines such as *Freibeuter, Aesthetic und Kommunikation, Prokla, Das Argument* and *Leviathan* from the 1970s. Sixty-eighters formed the core of the urban planning movement, which has had an influence in Berlin unmatched in other Western cities. From the early 1970s they played a key role in the Internationale Bauaustellung and gave rise to magazines such as a *Arch+* (Eichstadt 1990; Kühnert 1990). Those a decade younger, and with formative years in punk, give Berlin two weeklies of unusual quality, infinitely higher quality than say London's *Time Out* and *City Limits*, in *Zitty* and *Tip*. Berlin was the city of residence for godfathers (grandfathers?) of punk, Lou Reed, David Bowie and Iggy Pop. It is the only place outside of Manhattan which will guarantee large crowds of cognoscenti even today for the likes of John Cale and The Ramones. The post-punk ambience still dominates large sections of Berlin—including the twenty-year-old anarchists often from proletarian background who are so numerous as to make up almost an ethnic group in Kreuzberg, as well as the whole dressed-in-black, gothicized *Szene* whose contemporary heroes include the likes of Australian poet/rock star Nick Cave.

Berlin, in the early 1980s, then, was a space-time compressed, literally emptied-out island, deprived largely of economics and institutionalized politics with a hypertrophied culture. It was characterized at the same time by an overwhelming *dis*accumulation of capital or material goods and an *over*accumulation of cultural goods and was deliberately

[9] Central city area.

set up as not just a consumer, but as a *cultural* showcase from the West to the East, though this culture had largely fallen into the hands of those almost exclusively active in the sphere of culture–i.e., detested left-wing political and aesthetic avant-gardes. Thus Michael Rutschky (1987), essayist, photographer, editor of the magazine *Der Alltag* (subtitled *Die Sensation des Gewöhnlichen*), could remark that Berlin was all symbol, all image; it was indeed Baudrillard's simulacrum in the flesh.

New Genesis: Recasting the class structure

A very useful way to understand the transmogrified urban form that Berlin will come to have in the next decade or two is via consideration of its changing profile in social stratification. Nicos Poulantzas (1974) in *Les Classes Sociales dans le Capitalisme Aujourd'hui* made in this context a benchmark distinction in contrasting social class as a characteristic, on the one hand, of people or social agents and, on the other, of occupational places or social structure. This helps make some sense of impending change in Berlin. Only Poulantzas overestimated the importance of structures or places as a causal determinant, whereas, in fact, in Berlin it will probably be as much the new mix of social agents who will attract the new occupational structures as vice versa.

What about the structures or occupational places? At first in the light of impending reunification there seemed to be virtually unbounded optimism about Berlin's economic future. All the above-mentioned economic structures which had outmigrated, it was thought, would come right back. Berlin would be so strong economically as an urban center that the relocation of the federal capital from Bonn would be unnecessary. It was better to leave the economic force of being *Hauptstadt* back in Bonn. But as Niklas Kühnert (1990) observes, there is an enormous inertia in economic spatial relocation. That is, for example, banks will open much more substantial facilities in Berlin, but the finance center will remain in Frankfurt. Some corporations will relocate head offices from Düsseldorf to Berlin. But more common will be the opening of sub-head offices. Yes, SFB (Sender Freies Berlin) will expand–is indeed already taking on new functions and disposing of additional resources–as a locus for television production. But it will be dwarfed by West-, Nord- and Suddeutsche Rundfunk for a long time and will probably wind up as only one pole of a polycentric structure. In the mid-1980s Berlin's Senator for *Stadtentwicklung und Umweltschutz* (Urban Development and Environmental Protection) as well as its critics on the moderate left had ideas of Berlin developing as a high-tech center, given the weight of the Technical University. But few speak of this sort of development today.

This is partly because of infrastructure. West Berlin is still serviced by its intimate and famously 'quaint' airport, in comparison to the megastructures on the ground in Frankfurt and Düsseldorf/Cologne. The Federal Republic's rail system was notoriously deficient because what was, in the old Germany, an essentially east-west system was forced to run north and south. And anyone travelling to Berlin from West Germany in the past bore witness to the much worse East-West connections of the East German Reichsban. Connections of Berlin not just with the West but with Eastern metropoles like Leipzig and Dresden are at a perhaps 1930s standard of other Western countries. The same can be said for road and communications networks. More than one builder has been known to remark how the Soviets and the initiativeless East Germans just used up the '*Substanz*'[10] of the country's previously existing buildings and infrastructure.

What most commentators see as most likely is largely a city oriented around services (*Dienstleistung*). First, the national capital probably *will* relocate in Berlin, creating an enormous number of public-sector jobs, with very substantial multiplier effects on private-sector services. Second, the setting up of sub-head offices and of Berlin as a 'sub-center' for corporations, for finance, for the media and, inevitably, for the internationally expanding business sector itself will create another large swathe of professional-managerial service jobs. There is talk of developing a *La Défense*-like business center on the outskirts of the city. These new public- and private-sector jobs combined with what is already a very large culture services sector will provide the occupational places to form the uppermost class in what is likely to be a three-class system of stratification. Providing the people for these occupational places should not prove to be difficult. First, the inflow of students and other young people from Western Germany to Berlin has accelerated through the 1980s. West Berlin was an aging city up through the early 1980s, declining in population, and the rate of inflow of young people from the West was significantly lower than the death rates of the senior generation of aging native West Berliners. This trend was reversed from the mid-1980s (Profé, 1990). From this point in time West Berlin reversed its trend towards both aging and population decline, as the inflow of the young from the West for the first time effectively counteracted the death rate of the older generation. And this rate of inflow of students and other prospective professional-managerial employees has been only encouraged by reunification. The only major countertendency to this is that of East Berlin professionals—doctors, lawyers, symphony musicians—leaving Berlin altogether for jobs in western Germany (Kühnert, 1990).

The second major class in this schemata will be a class of workers, mostly manual, in industry together with skilled and semi-skilled industrial workers. Here a number of phenomena, it is likely, will continue to proceed simultaneously. The exodus of young,

[10] *Bausubstanz*: Literally building substance, but it refers to the 'core' of a building.

skilled male manual labor from eastern Germany is mostly not to Berlin but to western Germany. On the supply side of labor, this is largely due to the massive shutdown of the internationally uncompetitive eastern German factories. This crisis is, of course, made much worse by the drastic decline of orders from the other former Comecon countries for eastern German industrial goods. Eastern Germany, given the at least short-term hesitancy of German and other Western entrepreneurs to invest in their industry, is in danger of becoming a sort of Nordic *Mezzogiorno*, an underdeveloped half of the country supplying labor to the factories of the other half. And, indeed, historically the eastern part of Germany has been in virtually all respects less modern than the western part.

On the demand side, the opening up of the east as a market has and will continue to provide relatively quite favorable growth rates, hence employment prospects, for western German industry. But this is rather western German then western Berlin industry. Though western Berlin too has a significant number of industrial workers, these tend to be in lower-value-added and lower-skill areas than in western Germany. Where this decline in working-class jobs will be counteracted will be in the building industry, which will boom in Berlin for probably at least two decades, as the population of the city is expected to expand by some 33–50 percent. So both residences and office buildings will go up at a rate probably unmatched in other Western cities. Berlin is already a job-rich haven for otherwise unemployed architects from Paris to Dublin. Apart from East Berlin and East German building workers, there is already a very large influx of non-university graduate white-collar employees in Berlin, of technicians, physiotherapists, bank employees and the like, many of whom live in East Berlin and work in the West.

The third and bottom class of emergent Berlin, the 'sweated proletariat,' is not a universal phenomenon in today's advanced societies. It is characteristic of only one of what Gösta Esping-Andersen (1990) calls the 'three worlds of welfare capitalism.' In one of these worlds, typified by the Scandinavian countries, the existence of a strong welfare state provides a net which saves citizens from having to work for the poor wages of the sweated proletariat. In that Scandinavian welfare states are providers of welfare services, they also provide jobs for the lower ecehelons of society as employees who are working in and for the welfare state. In the second world of welfare capitalism, typified by Germany, there is also a strong welfare state, whose *Netz*[11] saves citizens from falling into the sweated proletariat. But inasmuch as the German welfare state is typically a *purchaser* of welfare services, it does not provide employment for high numbers of individuals. This, together with the strong presence of high-value-added work in the German economy, has tended to foster high levels of unemployment as well as a low female labor force participation rate. Germany thus has a very large manual working

[11] *Netz*: Welfare net.

class, say some 35–40 percent of the labor force, and a fairly small sweated proletariat.

The third world of welfare capitalism is typified in the Anglo-American cases. Here the welfare net was weak and/or significantly eroded in the 1980s. This fostered the development of a large sweated proletariat in the services and particularly in the garment industry as well as rather lower levels of unemployment than in Germany. Further, in the Anglo-American world only about 20 percent of the labor force is in the industrial working class, and the proportion of professional-managerial employees is especially high, matching levels of the manual working class. In the U.S. and the U.K. it is largely immigrants and minorities who make up the bulk of the sweated proletariat.

The point of this is that Berlin is likely to follow the Anglo-American model more than the German one. Alongside a high proportion of professional-managerial workers and a lower one of industrial workers, Berlin will probably develop a sweated proletariat on a significant level. This will be due to the influx of cheap labor that will crowd into the city. Germany, in the decades spanning the turn of the twenty-first century, is facing the prospect of waves of immigration from the East at about the rate the USA experienced in the decades spanning the turn of the twentieth. In the next 25 years this could come to as many as four or five million new immigrants. Germany will probably be able to accommodate this because of long-term prospects of a thriving economy and because the birth rate—of 1.4 to 1.5 per female—is well below the 2.2 replacement rate and indeed is probably the lowest in the world.

But in the short term this means a quick and huge influx of Poles, Jews and especially Russians, many of the whom do not have average German levels of skills. The federal government tends to control this and distribute incomers proportionally among West and East German locales. But the very scale of immigration will be impossible to control. First,there will be enormous numbers of illegal immigrants. Second, very large proportions will be able to avoid relocation controls (Hoffman-Axthelm 1990). And third, a great proportion of both of these will settle in Berlin. It is likely that the level of the welfare net will decline somewhat to help pay the costs of reunification. Further, many of Berlin's illegal immigrants will be unable to claim benefits. Finally, Berlin will attract a large number of former East Germans who do not have easily classifiable skills. With these newcomers, facing much higher wages than they previously made in Eastern Europe, are present all the ingredients of the recipe for the rapid building of a sweated proletariat. Southern California in the 1980s built a sweated proletariat on the backs of illegal Mexican immigrants (Davis, 1990). *Berlin's* 'Mexicans' of the 1990s and the first decade of the twentieth century will come from Russia and from Poland.

But Berlin already had quite a potent ethnic mix. Indeed, most major German cities have populations, some 10 percent or more, comprising Mediterranean peoples, mostly Italians, Yugoslavs and Turks. The problems involved were visible in microcosm at a concert given by the politicized New York rap group Public Enemy in November 1990 in Hasenheide in Berlin Kreuzberg. Outside the concert Turkish youths from a Maoist organization were handing out leaflets protesting against a demonstration in East Berlin. A Leipzig police officer had shot and killed a young soccer fan the previous week. This soccer fan was identified with racist, skinhead hooligan elements in the East. Indeed, the leadership of the West German Neo-Nazi party had been active in recruiting East German skinheads, who had been carrying out racist attacks on Vietnamese *Gastarbeiter* and other minorities in the East. A week after the shooting, young East Berlin soccer fans, including large numbers of 'skins,' marched in protest against the shooting. The Turkish Maoists protesting outside the Public Enemy concert were against the demonstration.

Inside the auditorium was a mix of Turkish, black and white youths decked out in head bandanas, gang jackets and the whole rap regalia. I tried to stick close, for safety's sake, to the older and respectable Irish architects whom I met in the ticket queue. The rappers, Public Enemy, did not really know how to address the audience, all of their celebration of the end of the Wall and reunification being greeted by a marked lack of enthusiasm. I left the concert early, having to catch an early morning flight the next day for Paris. In the *Tageszeitung* the next morning, though, it was reported that 112 weapons were found among the audience in the body searches preceding the concert. Also, about midnight open fighting broke out between gangs of Turks, blacks and whites.

Despite all the media treatment and worry of public opinion about racism among East German youth (Schroeder, 1990), the extreme right Republikaner party fared very poorly in the first all-German national elections since 1932 of December 2, 1990. The Republikaner failed to secure enough votes to get into the Bundestag at all, doing particularly poorly in the German DDR. In Berlin, where they had gained some 7–8 percent of the vote in the April 1989 contest, they dropped to some 3 percent. Indeed, perhaps with reason the German media tend to regard Germany as a beacon of Western reason in comparison to the what they see as the rampant nationalism and racism of Polish, Austrian and Russian neighbors. It is interesting that the now mass right-wing party, the FPÖ, in Austria backs unification with Germany, while no significant political force in Germany itself wants a replay of Anschluss.

Yet the dangers of a resurgent, though doubtless milder, fascism must not be neglected. Alone in Germany, in comparison with France, England and America, is national identity so closely bound up with *ethnic* identity. Indeed, it has been argued that a decoupling of the two is necessary for any sort of democratic and plural polity in a Germany, whose cities may contain over 20 percent minorities twenty years hence. Alain Touraine (1990) has recently argued that the only significant social movements of the present are of ethnicity and nationalisms. He has pointed out that whereas Britain tends to understand minority ethnicities as structured in communities, in France the less substantial existence of communities means that minorities must be unmediatedly integrated as national citizens into the French state. The German case, with its understanding of national identity in a racial sense, poses yet different problems.

Politics of urban change

Architects and urban planners in Berlin under the age of fifty have been schooled in reaction to the excesses and anti-urbanism of the Garden City idea and the hyperrationalism of Corbusier. Anathema to them are huge council houses, unused public space and urban and suburban sprawl. In this sense they have been similar to the Jane Jacobs-inspired 'anti-development' lobby in other Western countries. But urban utopians in Germany, with its strong and long-time presence of a significant environmental movements, have taken two forms. One of these is indeed green. This was the official reigning ideology of the Red-Green coalition in power in Berlin from April 1989 through December 1990, whose principal spokesperson in government was perhaps Michaele Schreyer, Senator for Urban Development and Environmental Protection.

Schreyer (1990), herself a social scientist (an economist), outlined the main features of this green urbanism. First, she said, we must 'start where we are,' i.e., no modernist blueprints imposed abstractly from without. Second, she opposes urban sprawl with Berlin extending far out into the suburbs on the model of Los Angeles or London. Third, Schreyer–in counterpoint to her critics–does hold that Berlin has room for growth. And finally she and her advisor Béate Profé (1990) propose that there be very broad spaces of green park right through Berlin's inner city, proposing somewhat on the model of New York's Central Park that this green urban space will be significantly used by the public. What the Greens propose is what they call a *Stern* conception, in which from Berlin's center a multiple-pointed star will radiate to the S-Bahn ring, which is about eight miles in diameter and encircles the center of the city. On this model there will be green park between the 'fingers' of the star, which will reach a series of points contiguous to the various stations on the S-Bahn ring, around which will develop shopping and general service centers.

Wulf Eichstadt, formerly Kreuzberg and now Charlottenburg architect and planner and formerly a key figure in the IBA, fills this conception out. For Eichstadt (1990) the historic division of Berlin into two is something that should not be erased, and to do so would be to fall into the modernist fallacy of travestying real historical identities. He proposes that the eastern Innenstadt–of Unter den Linden, Leipziger Platz and Potsdamer Platz–be developed to form a sort of natural point of attraction for shopping and culture of East Berlin areas of Friedrichshain and Prenzlauerberg. Eichstadt further supports a polycentric structure for the whole of Berlin with 'mini' city centers in Schöneberg, in Kreuzberg and so on.

If this green urbanism is one contemporary political critique of the Garden City, high-rise blueprint modernism, the other is closer to Anglo-American critics of modernist 'urbacide' such as Jane Jacobs and Marshall Berman. German urbanists commonly refer to this as a 'Verdichtungsidee,' i.e., a high-density conception. Crucial proponents of this sort of notion are Dieter Hoffmann-Axthelm and the Gruppe 9. Dezember (December 9th Group). The December 9th Group proposes low-rise, high-density, multi-purpose apartment buildings in the whole area inside of the S-Bahn ring, much on the model of the classical Berlin (late 19th century) *Gründerjahre* buildings with courtyard and back courtyards. The idea is opposed to zoning and proposes an urban complexity in which small shops will form much of the ground floor of buildings (as they often do even now) on residential streets, but in which service offices and artisan workshops will also exist in buildings side by side with apartment dwellers (Charta Mitte Berlin 1990).

The high-density utopia would call for a Berlin on very much the unified model of the pre-Nazi period, in which the Potsdamer Strasse was effectively the connecting link between western and eastern city centers. Unfortunately, architect Hans Scharoun, seemingly deliberately, sat the imposing (though possessed with the most beautiful and functional of interiors) bulk of the Staatsbibliothek[12] right in the middle of the old Potsdamer Strasse, winding a new street around the building which no longer connects properly to Potsdamer Platz. The utopia of the 9. Dezember Gruppe would be for pulling down the Staatsbibliothek, and rebuilding the old link. Hoffmann-Axthelm (1990) holds that Berlin can indeed take on up to 5 million population. He argues that the Green urbanism outlined above would only lead to the same old excesses of Garden City modernism, unused public space and unsightly urban sprawl. He notes that the anti-zoning, high-density alternative would bring jobs and residences in proximity to overcome transport problems.

It further would provide an infrastructural focus of clubs, cafes and sports and cultural institutions for otherwise underregulated urban youth. Both East Berlin youth and the expected new immigrants from Russia and Poland will be particularly susceptible to deficits of 'moral regulation.' In East Europe, Communist erosion of other localized instances of such regulation left only the state as an effective instance of regulation. With its destruction, deficits of social control are sure to follow. This goes doubly for youth from Poland and Russia, who in addition will be uprooted and living in a foreign ambience. The high-density solution is much more amenable to moral regulation than the isolated council housing in the city's suburbs. If the French experience is anything to go by, in the 12 November 1990 Paris riots the majority of those arrested for the violence and looting adjoining the national demonstration of high school students were black and

[12] State library.

Arab youths operating in gangs who lived in high-rise council housing in the Paris suburbs (*Libération*, 1990).

But how much do the utopian politics of either Greens or the high-density urbanists have to do with *Realpolitik*? In the all-German and Berlin elections of 2 December 1990, the left and especially the Greens took an unexpected trouncing. The West German Greens fell from 8 percent in the 1986 elections to below the 5 percent level and hence now have no deputies at all in the Bundestag. Analysts blamed it on their desertion of the ecology issue to constitute an alternative to the left of the SPD.[13] On the federal level Oskar Lafontaine, deserted by the SPD hierarchy, took an even worse-than-expected drubbing. But the biggest surprise was the defeat of the SPD and Walter Momper in Berlin. Momper had become the Berlin and all-German media darling. He was the Oberbürgermeister[14] of reunification, living simply in Kreuzberg, touted as their next SPD Kanzler (or at least SPD leader) in the tradition of former West Berlin mayors such as Willy Brandt and Hans Jochen Vogel.

The defeat of Walter Momper seemed to follow directly on his forcible eviction of, and subsequent violent struggles of the police with, squatters in the Mainzer Strasse in East Berlin's Friedrichshain. The squatters, especially those engaged in violence, were in great majority West Berliners. Often from Kreuzberg, a number of them black-cladded anarchist *Autonomen*, the logic on their side was the incredible shortage (some 200,000) of apartments in West Berlin, while some 30,000 flats in East Berlin stood empty. Their forcible eviction was a patent electoral ploy by Momper in order to win law-and-order conservative voters. Momper knew that the effect of the forcible eviction off the *Chaoten* was likely to split the Red-Green coalition (Schweitzer 1990). For him all the fighting and struggles that had plagued the entire short lifetime of the coalition had in any event made government unworkable in Berlin. He also knew that he would have to govern in a Great Coalition with the Christian Democrats (CDU). Pundits and even the CDU predicted that Momper would receive some 40 percent of the vote and the CDU something over 30 percent. In the event, the figures were reversed. Particularly surprising was the strong support in West Berlin for the CDU and the weakness of the left in West Berlin. The CDU, who were so sure of defeat they did not plan a celebration party, have been faced with the unexpected task of being the powerful effective governing party in a great coalition. And Momper is reviled by all, on left and right, his political stock having fallen more quickly than Donald Trump's fortune.

How can the debacle be explained? First, with the specter of instability with reunification, the prospective inundation of Berlin by eastern Europeans (The 'Russians

[13] Social Democratic Party.

[14] Mayor of the entire city of Berlin.

are Coming' specter) and the accompanying financial insecurity, voters opted for stability and not for the crisis-ridden bickering of the Red-Green coalition. Second, there may well have been very important concern of voters for the existence of a liberal *Rechtsstaat*,[15] i.e., of a democratic and plural regime. This might explain the surprising successes of the Free Democrats both in the West and especially in the East and the demise of the Greens on the left and of the Republicans and the CSU[16] on the right. The SPD-oriented *Die Zeit*, published by Helmut Schmidt, condemned the Mainzer Strasse squatters and the political operations of the Greens in this connection on the basis of a liberal Rechtsstaat (Schueler, 1990).

Perhaps this *rechtsstaatlich* belief illuminates the effective coldness of left and left center intellectuals to Lafontaine's candidacy, which stressed the *economic* costs of reunification. Surely this had an appealing anti-nationalist ring to it. But if one reads it back into Lafontaine's and the left's attitudes of early autumn 1989, it neglects the liberal and *rechtsstaatlich* inclinations and opportunities of a whole nation of East Germans deprived of the right to vote, to travel and to free expression. Hence left debates have shifted to that very *rechtsstaatlich* issue of the possibility of a new constitution for a reunified Germany. And social scientists like Ulrich Preuss have become involved in the debate, calling for a new constitution for a now-unified Germany in place of the Federal Republic's externally imposed constitution. Perhaps the important point, however, stands perpendicular to this debate and is precisely the very extent of legitimacy and public belief in the validity of the *present* constitution. Many nations have quite impeccable constitutions without any real weight at all. But it is likely that outside of the United States no nation has the sort of psychological investment in its constitution as in Germany.

The point, moreover, is that the *left* is coming to see the constitution and *Rechtsstaatlichkeit* as the key issue—as an alternative and supplementary 'post-materialist' value to the ecological issue. This runs parallel in important senses to what left intellectuals in other countries have understood under headings of 'civil society,' 'difference' and 'complexity.' In Germany convergent notions have been developed by sociologist and environmental commentator Ulrich Beck with his concept of 'reflexive modernization.' For Beck (1986) structural change undergone along with the process of modernization forces individuals to be increasingly 'reflexive'—that is, increasingly to make decisions, to choose between alternatives. Unleashed from the tutelage of traditional structures (state, family, church), individuals on this view also must increasingly take

[15] *Rechtsstaat*: Literally 'state of law,' meaning state founded on a legal, constitutional basis which at the same time guarantees legality for individuals. It usually entails parliamentary sovereignty, political party pluralism, the hegemony of civil society and a bureaucracy using general rules applicable to all.

[16] Christian Social Union.

responsibility for the consequences of their action. The political directive implicit in this is the creation of structures which would further enhance the sphere of reflexive and autonomous action.

Complementary to these theses is the idea of '*Kulturgesellschaft*'[17] developed by Social Democratic analysts such as Peter Glotz, Hermann Schwengel (1988) and Eberhard Knödler-Bunte (1990). On this account modernization is a process of structural differentiation, and in advanced societies a third sphere will differentiate out from that of state and economy—i.e., a realm of culture. Such differentiation has proceeded most pronouncedly in those societies in which the role of heavy industry and the weight of the industrial working class are declining. This growing third sphere encourages the displacement of politics from production to consumption, from the material to the cultural. In such a third sphere, which other analysts (e.g., Urry, 1981) would liken to Gramscian 'civil society,' post-materialist principles of justice and 'difference' are pervasive. Gramsci himself understood, in counterposition to the economy and material needs, 'culture' and 'subjectivity' almost as interchangeable terms. And the reflexivity of agency or subjectivity on structure is in this sense not vastly different than the reflexivity of culture itself on the state and economy (see Luhmann, 1984). In any event, with an international left notoriously short on ideas and Western societies having undergone conservative routes to modernization, some on the German left seem to be developing a new battery of political concepts.

Conclusions

Whither Berlin? What sort of urban paradigm will provide the lines for its imminent breakneck development? The green urban utopia seems to be precluded as the election results of December 1990 spoke for the commitment of East (and West) Germans to their automobiles, which they thought threatened by Green pronouncements. Further, the middle mass of voters were put off by what they saw as the unrealistic nature of (what little there was of) Green economic analysis. But equally the high-density urban utopia—which would restore the dimensions of Simmel's and Benjamin's Berlin—has little chance of realization. Where would one find all the enterprising property developers to build all of these structures? The immediate housing shortage in Berlin is so drastic that government will have little choice but to build as fast as they can and—reluctantly—to a substantial extent through public institutions. Pressures of the moment will make worked-out planning very difficult. And the most likely outcome is that of urban sprawl, with

[17] *Kulturgesellschaft*: Literally, culture society, or society in which the cultural sphere is predominant.

suburban high-rise public housing, and all its attendant social problems, dotting the landscape *out*side of the S-Bahn ring.

There is an irony in Berlin politics. In this last local election, the right of the political spectrum–Republicans and right wing of the Christian Democrats–wanted a Berlin that had the dimensions and status of a Weltstadt, while at the same time putting limits on ethnic diversity, whereas the Greens' proposed multi-culturalism was combined with a low-growth perspective that would have prevented Berlin from ever becoming a Weltstadt. The real irony of Berlin politics, however, is that no political force will be able to put much of an imprint on the shape of things, which will be decided external to politics by overwhelming economic and especially demographic pressures. Politicians in power will above all be engaged in crisis management, forced not to act but react. Simmel and Benjamin's modernity in Berlin was that of an emergent large-scale industrial capitalism. Today's Berlin is in rather different times, captured by Ulrich Beck as '*another* modernity'. Perhaps the business of politics, and left politics (can we still meaningfully talk of 'left' politics?) in such a contemporary modernity is one of acting within limits and with the limited objective of creating and leaving maximal space for civil society; for reflexivity, complexity and difference.

It has now become almost conventional to speak of a changing locus of urban modernity–to treat the modern, that is, the paradigmatic space of urban *restlessness*, as relocating from nineteenth century Paris via a brief stopover in fin-de-siècle Vienna to Berlin in the aftermath of World War I and then to speak of a long hegemony of New York–from the bebop and abstract expressionist-inflected middle forties to the crisis of hegemony the early 1980s. A number of American analysts at first thought that New York's hegemony would pass to Los Angeles, home of international popular culture, budding art museums, telescoping universities and a characteristic and influential 'designed (i.e., from urban space to dress styles) environment.' These commentators should spend some time in Berlin, as, it is likely, will many of the Simmels and Bejamins of twenty-first century modernity.

References

Beck, U. 1986 *Risikogesellschaft*. Frankfurt: Suhrkamp. Trans. *Risk Society*. London: Sage, 1992.

Charta Mitte Berlin. 1990 Gruppe 9. Dezember, July 1990.

Davis, M. 1990 *City of Quartz*. London: Verso.

Eichstadt, W. 1990 Architect and planner, leading figure in Internationale Bauaustellung. Interview, Berlin, 29 November 1990.

Esping-Andersen, G. 1990 *The Three Worlds of Welfare Capitalism*, Cambridge, England: Polity.

Frisby, D. 1985 *Fragments of Modernity*. Cambridge: Polity.

Gerschenkron, A. 1962 *Economic Backwardness in Perspective*. Cambridge, Mass.: Harvard University Press.

Giddens, A. 1984 *The Constitution of Society*. Cambridge: Polity.

Giddens, A. 1990 *The Consequences of Modernity*. Cambridge: Polity.

Hoffmann-Axthelm, D. 1990 Berlin urban historian; Participant Internationale Bauaustellung; leader of 9. Dezember Gruppe. Interview, Berlin, 13 December 1990.

Knödler-Bunte, E. 1990 Director, Potsdam Kolleg für Kultur und Wirtschaft, Interview, Berlin, 19 December 1990.

Kühnert, N. 1990 Architect and planner; editor of *Arch+*; Interview, Berlin, 19 November 1990.

Lash, S. 1990 *Sociology of Postmodernism*. London: Routledge.

***Libération*.** 1990 13 November, Paris.

Luhmann, N. 1984 *Social Differentiation*. Cambridge: Polity.

Poulantzas, N. 1974 *Les Classes Sociales dans le Capitalisme Aujourd'hui*. Paris: Seuil/Maspero.

Profé, B. 1990 Referentin to M. Schreyer, Senator für Stadtentwicklung und Umweltschutz. Interview, Berlin, 23 November 1990.

Rutschky, M. 1987 'Statt zu arbeiten,' *Der Alltag*, nr. 2/87, pp. 120–135.

Schreyer, M. 1990 Senatorin für Stadtentwicklung und Umweltschutz. Interview, Berlin, Rathaus Schoeneberg, 23 November 1990.

Schroeder, B. 1990 'Extreme Rechte vor der Wahl,' *Zitty*, 24/90, pp. 16–21.

Schueler, H. 1990 'Kein Frieden im neuen Land,' *Die Zeit*, 23 November 1990, p. 1.

Schweitzer, E. 1990 'Hausbesetzer in Ostberlin,' *Zitty*, 24/90, pp. 8–11.

Schwengel, H. 1988 *Der kleine Leviathan*. Frankfurt: Athenäum.

Touraine, A. 1990 *Die Tagezeitung*, 10 November, Berlin.

Urry, J. 1981 *The Anatomy of Capitalist Society*. London: Macmillan.

Ten

CAN THERE BE A POSTMODERNISM OF RESISTANCE IN THE URBAN LANDSCAPE?

David Ley and Caroline Mills

University of British Columbia *College of St. Paul and St. Mary*

The Biennale Exhibition, 'The Presence of the Past,' held in Venice in 1980, became in some quarters an international manifesto of postmodern design, the first large collaborative display of a movement growing in self-confidence. It also became the immediate target of Jürgen Habermas (1983) when he delivered his defense of modernity to the citizens of Frankfurt two months later. The defensive lines were drawn sharply: the Biennale was a presentation of 'Antimodernity,' and from there Habermas could project postmodernism onto neoconservatism, attempting, through a political critique, to rescue the modern project. This transposition was not missed by Paolo Portoghesi, one of the organizers of the Biennale. Claiming, paradoxically, a 'continuity with the great tradition of modern art,' Portoghesi, in a thinly veiled reference to Habermas, rejected 'the new conservatives . . . [who] speak of an incomplete project of modernity that must be continued' (Portoghesi, 1982, p. 8).

How quickly a discourse on design has become a discourse on politics and, indeed, has turned to critical imputations of personal politics! It is also instructive to note how speedily the employment of partisan rhetoric stifles serious discussion, for, as we hope to show, the projects of Habermas and Portoghesi have a good deal in common. In his Frankfurt address, for example, Habermas set himself at some distance from the modern avant-garde in his advocacy of the lifeworld and its vital cultural heritages while still rejecting 'mere traditionalism.' When he returned to the problem of modern architecture in later papers, he tended to agree with its critics that after 1945 it became a 'disaster' (Habermas, 1989b, p. 41), its products 'monstrosities' (1989a, p. 8), although this observation is camouflaged beneath renewed criticism of neoconservatism and a defense of the modern movement, which we shall consider later.

The apparent convergence between Habermas and Portoghesi goes beyond a critique of modern architecture. Portoghesi urges a deflection of influence and power away from entrenched elites toward participation with a broader public. The sharing of power is to be matched by shared frames of meaning. On the one hand, postmodern design maintains continuity with modern design in generating a set of meanings accessible only to a speech community of architects; but on the other hand, unlike modernism, it is double coded, engaging self-consciously with the popular realm, by re-presenting meanings which are integral to local cultures. It is above all 'an architecture of communication' (Portoghesi, 1982, p. 11). This communicative orientation evokes the ideal speech situation, the formal model repeatedly advanced by Habermas as the vehicle to correct the unequal relations between system and lifeworld and to inaugurate democratic exchanges for collaborative decision-making. These common objectives suggest opportunity for recognition of a common project between Habermas and Portoghesi, but no such identification occurs. As the combatants assume political postures, each strives to establish a position beyond the left flank of the other. Rhetorical flourishes replace careful engagement and any opportunity for a shared task is lost (cf. Rorty, 1985).

Indeed the first task in this chapter is to clear the ground through a literature that is characterized by a high level of polemic on all sides. But it is not polemic alone that has impeded serious discussion of postmodernism. Definitional inconsistencies abound. Moreover, the terms of reference brought to the debate have often been flawed, so that criticism has been mired by the same façadism that it has condemned. We argue that the discussion must include not only cultural forms, like buildings or artwork, but also the cultural and political practices implicit within them. Postmodernism is not a mere morphology of artifacts, but also the practices and activities by which artifacts are constructed. Such processes, although they may end in commodification–though this term, itself, is not unproblematic–encompass singular moments where it is possible to open up spaces for opposition, to conduct rituals of resistance. Indeed, the very notion of art being coopted in the interests of capital implies a subversive possibility that must be headed off. Second, then, we seek to uncover a point where it might be possible 'to unlock the critical moment in postmodernism' (Huyssen, 1984, p. 9). The chapter ends by adding regional specificity to the argument in a brief review of landscapes created in Vancouver's inner neighborhoods over the past twenty years. Only then do we feel able to offer a more tangible answer to the question that motivates this chapter: can there be a postmodernism of resistance in the urban environment?

Rats, posts and other pests . . .

But first there is a minefield of spent charges to negotiate. The paradigmatic case of postmodern art is architecture, but within the architectural profession the reaction against postmodern design from established figures has frequently been vigorous. In 1981 Aldo van Eyck's Annual Discourse to the Royal Institute of British Architects under the title of 'Rats, Posts and Other Pests' set the scene (and the level) for discussion in the early 1980s (Jencks, 1986, p. 11). It was soon to be followed by appellations of 'transvestite architecture,' 'Charles Junk' (sic), and a 1983 debate generating such condemnations as 'ephemeral, a throw-away, subject to the caprice of commercial culture and the exigencies of the marketplace' or 'neofashion for the bored, the rich, the jaded, the blind' (AIA, 1983). As Christian Norberg-Schulz was to observe later, the vigor of these reprimands suggested that postmodernism was too important to ignore, its criticisms too central for an understated response. But the response, to the extent that it was substantive, was primarily aesthetic, concerned with the retention of what a corpus of professionals defined as good urban form.

In the social sciences as well, a common assessment of postmodern culture has been critical. An arsenal of conceptual hardware has been assembled: the 'postmodern city' is assailed as a place of spectacle and surveillance, where aestheticisation and commodification lull the consumer into a quietism before the social control of a market-driven hegemony. This, of course, is the barest of skeletons, and arguments are assembled upon it in sometimes elegant and persuasive forms. There is, moreover, undoubtedly some truth to such a depiction. The important question then becomes, how complete is this truth, how adequate is the argument, both theoretically and empirically? Several observations appear relevant.

First, the imagery of the postmodern city employed by its critics is one of social control: superficial, garish, tasteless, deceitful and manipulative. It is notable how closely this vocabulary is shared with the Frankfurt School's condemnation of the culture industry, where the 'reified false consciousness of industrialized mass culture has settled like a pall over history' (Brantlinger, 1983, p. 226). The protests of Horkheimer and Adorno a half-century ago as they contemplated cultural life in the *modern* city have a very current ring to them. A generic culture of 'ready-made clichés to be slotted in anywhere' created an alienated consumer who 'becomes the ideology of the pleasure industry, whose institutions he cannot escape' (Horkheimer and Adorno, 1972, pp. 154, 158). With little qualification these arguments are carried forward by influential authors like Debord (1973), Foster (1985) and Harvey (1989). Despite the sea change that we are told has led to the postmodern condition, a peculiarly modern theoretical model is sustained in its interpretation. Moreover, the discourse of the culture industry portrays an unusually claustrophobic world, where a homogeneous and tightly controlled public culture is projected unproblematically upon a passive citizenry.

This structural model has been criticized often enough before—not least in the realm of cultural geography—but some particular shortcomings bear repeating. Such models of hegemonic control present the consciousness of the masses as monolithic and unproblematic, passive and without the potential for resistance. The view of mass culture is distant and elitist. Soja's (1989) view of the surveillant state in Los Angeles, for example, is a view from on high; as the noose of total social control is drawn tightly around the city, we do not know if any member of the thousands of cultural worlds in that city has noticed, for no other voice or values are admitted other than those of the author. As in so much of the literature on cultural hegemony, the social control of consciousness is alleged but never proven. When we look for the voices of the manipulated masses we encounter in the text a gaping silence, a silence that encourages the disturbing thought that perhaps the manipulation of mass consciousness is accomplished also by the theorist, who as *spectator* him/herself posits the existence of the spectacle as a source of confusion only for the undifferentiated Other. We might ask why the author-ial/-itarian I/eye does not take this criticism of a way of seeing in others more self-consciously (cf. Deutsche, 1991).

Second, and related, much of the discussion around postmodern design is characterized by what may be called a façadism which does not problematize the meaning of a building or a landscape. Totalizing views of a culture read off its impact upon an artifact under the assumption that the artifact expresses and reproduces mechanically the social relations imputed to the culture. This was the complaint directed against an earlier cultural geography which, in less sophisticated ways, treated dwellings as mere forms, carriers of an undifferentiated culture. Today, the dominant culture is represented as less harmonious and more conflict-strewn, but the treatment of landscape remains incomplete. Consider, for example, Harvey's (1990) reading of the redevelopment of Baltimore's Inner Harbor. While in important respects his interpretation rings true, it is rather too hermetically sealed, both theoretically and empirically, and its absences are as strategic as its presences. We accept his account of the remaking of the Inner Harbor and of the public investment that subsidized private development. Informed by his theoretical viewpoint, what has resulted is a 'carnival mask,' disguising the alienations of commodification, a quintessentially inauthentic 'postmodern' landscape. But are there not also important absences, a depthlessness, *to this account?* Does the landscape not fit too closely the theoretical garb cast over it? If we treat the Inner Harbor less as the projection of a theoretical position and more fully as a historical-geographical landscape, other questions come to mind. Harvey does not say what alternative uses for a decaying industrial landscape he might propose, other than to rue the passing of a city fair in the early 1970s. He does not allow that the carnival might generate outcomes which are concealed by the hermetic concept of the spectacle. Mundanely, but importantly, what are the multipliers in the local economy and the job creation associated with them?[1]

[1]One of us found informative the comment of a black Baltimore cab driver who expressed enthusiasm for the (continued...)

What has been the effect on the city's tax base and its capacity to offer social services? May even the carnival be able to sustain practices which escape the imputed social control of spectacle? Is it not possible, for example, that the development of social and family relations among visitors might permit the advancement of the values of the lifeworld rather than 'the impoverishment, the servitude and the negation of real life' (Debord 1973, paragraph 215; compare Ley and Olds, 1988)? These are all empirical questions to which we have no answer, but they are questions that must at least be asked and not concealed by the rhetoric of the spectacle.

Façadism, then, may lead to casual empirical study where buildings and landscapes are treated as conveniences for the outworking of a larger theoretical project. The impressive scope of Jameson's (1984a) reading of postmodern culture has perhaps limited commentary on some of the large assumptions that underlie his thesis. Jameson begins by observing that his own conception of postmodernism was initially formed by debates in architecture. This acknowledgement makes his selection and interpretation of the Hotel Bonaventure in Los Angeles of quite central importance to his argument;[2] it is an archetype of 'postmodernism and the city.' Here, in a world enclosed by a glass skin, with its breathtakingly confusing lobby across which people float on kinetic 'people-movers,' one enters the all-embracing grasp of 'postmodern hyperspace.' The alarming sense of social dislocation is heightened by one's experience when outside the building. Mirrored glass walls offer no sensation of depth, no reproach to the old city nor any promise of utopian transformation; they offer only reflections. 'This strange new surface . . . renders older systems of perception of the city somehow archaic and aimless, without offering another in their place' (1984a, p. 62). Jameson's architectural example stands for the 'postmodern' world as a whole: the dominance of image over substance, the mystification of representation, the inability of subjects to orient themselves in a stable social context (Wexler, 1990).

Geographers have keenly embraced Jameson's architectural example, perhaps because it codifies a belief that spatiality is the overriding concern of the postmodern world. But does the Bonaventure really illustrate a radical departure from modern urban space? Jencks (1986), Hutcheon (1986-87), Preziosi (1988) and others suggest that it does not. If modern architecture is distinguished by the value it places upon abstraction, truth to

[1](...continued)
Inner Harbor redevelopment because of the traffic it generated for him. It was this same promise of minority jobs and business opportunities that won the support of black community leaders for the project (Frieden and Sagalyn, 1989, p. 115).

[2]Notable here is the consistent misspelling of the Bonaventura (sic) throughout the article; also the reference to the architect 'Postman' (sic) on p. 83. For a severe critique of other aspects of Jameson's thesis, see Preziosi (1988).

materials and the purity of form, then design that carries to extremes this self-referentiality is *late* modern, *not* postmodern. This is an ultra-hermetic world, justified on the basis of internal relations rather than through connections to its social or physical setting. Dismissive of the past, of old rules and traditions, this is pure aesthetics, albeit aesthetics dedicated to a squarely economic function. Jencks argues that this disagreement over categorization is far from pedantic, for classification serves a larger theoretical purpose. Aside from the building's overall form (Jencks, 1981, p. 35), there are several clues that point to the Hotel Bonaventure's late-modern pedigree. The mirrorplate on the skin of the structure (Jameson's 'reflector sunglasses') was a common late-modern element in which the striving to minimalism was taken to its limit—the disappearing building. Second, the rejection of the street and the neighborhood which Jameson discerns contradicts the self-conscious contextualism of postmodernism.

Perhaps most important, the sense of hyperspace, vertigo and disorientation within the Bonaventure was, for commentators from Baudelaire to Simmel, a quintessential experience of the *modern* city (Chorney, 1990). In modern art, Kern (1983, p. 142) has noted the rendering of space by Cézanne and the cubists as a 'reduction of pictorial depth and the use of multiple perspectives,' akin to the depthlessness and shifting reflections noted by Jameson. Kern (1983, p. 179) also cites Nietzsche's famous depiction of the emptiness of modern space without tangible grounding, directionless and alienating: 'Whither are we moving? Away from all suns? Are we not plunging continually? Backward, sideward, forward, in all directions? Is there still any up or down? Are we not straying as through an infinite nothing. Do we not feel the breath of empty space?' Straying as through an infinite nothing is precisely the imagery of spatial perception conveyed by Jameson in the Bonaventure, where 'it is quite impossible to get your bearings' and where

> *Hanging streamers indeed suffuse this empty space in such a way as to distract systematically and deliberately from whatever form it might be supposed to have; while a constant busyness gives the feeling that emptiness here is absolutely packed, that it is an element within which you yourself are immersed, without any of that distance that formerly enabled the perception of perspective or volume.*
>
> (1984a, p. 83)

There are further clues that Jameson's treatment of postmodern culture bears much more affinity with the modern era. The effacement of the personal, of feeling and of subjectivity which Jameson locates in postmodern art is precisely the criticism directed against modernism by its critics as they urge a restoration of 'the imperatives of the spirit' (Jencks and Chaitkin, 1982, p. 217), of places which evoke redemptive social meanings. Such semantic loading commonly includes historical referents, but these are conspicuously absent from the Bonaventure, which displays in true modern form the ecstasy of the new.

What the Bonaventure *does* express is the national or multinational corporation, and for Jameson this is a more central property than any other. Indeed he acknowledges that the hotel 'is in many ways uncharacteristic' of other postmodern buildings. Why then has this eccentric case been selected? Because it can conveniently then be turned to support the principal thesis that postmodern design is 'grounded in the patronage of multinational business,' that 'postmodern culture is the internal and superstructural expression of a whole new wave of American military and economic domination throughout the world' (1984a, p. 57). The necessity of this conjunction is a necessity for a theory and a politics which require it: such a discourse disregards the identity of cultural phenomena and remakes them to a form consistent with its own essential categories (cf. Deutsche, 1991).

According to this view, a second essential property of the Bonaventure besides its ownership which qualifies it as an icon of postmodern culture is its age. Built in 1976, the hotel falls within the periodizing net which Jameson throws over the postmodern, a classification which he notes 'is both inspired and confirmed by Mandel's tripartite scheme' of capitalist development (1984a, p. 78). This imputed equivalence raises a host of problems. First, like Harvey's depiction of postmodern culture, it contains the usual shortcomings of base-superstructure arguments. Second, it is indiscriminate and totalizing, extending the identity of the postmodern to all phenomena which fall within its historical boundaries, even if, like the Bonaventure Hotel, they do not meet the terms of a more specific cultural definition. Such a portmanteau concept very quickly becomes meaningless. If all that cultural objects share in common is their historic era, then, in the absence of careful empirical specificity, all that gives them a theoretical unity is the blunt idealist vehicle of *Zeitgeist*, a shared spirit of the times. Third, Mandel's periodization does *not*, in fact, coincide with Jameson's. Mandel's phase of late capitalism, equated by Jameson with postmodernism, is dated from 1940. It thus includes the cultural period of high modernism as well as postmodernism. The portmanteau of postmodernism becomes ever fuller, ever more chaotic. Our confusion is complete when Soja, who also seems to accept this periodization, refers to Mandel's all too modern Marxian account as the work of a 'foundational postmodern geographer' (Soja, 1989, p. 61).

The political mandate which, as Jameson, Soja and Harvey freely acknowledge, infuses their own arguments does not adequately specify the precise links between the neoconservative politics of the United States and Britain in the past 15–20 years and postmodern cultural forms. It is worth comparing Mary McLeod's point of departure which acknowledges 'the difficulties of any simple equation between postmodernism and a political position' (McLeod, 1989, p. 23). This point has been made frequently before in the context of modernism, which was appropriated by both Fascism and Stalinism in the interwar period. No modern landscape makes the point more forcefully than Brasilia, 'planned by a left-center liberal, designed by a Communist, constructed by a developmentalist regime, and consolidated by a bureaucratic-authoritarian dictatorship' (Holston, 1989, p. 40). In this context it would be surprising to see an unambiguous relationship between political ideology and what is generically called postmodern culture. Huyssen (1984) notes that while postmodern culture has been challenged by the Left, it has also been criticized by neoconservative authors. Muddying the waters still further is the unequivocal criticism of modern design and planning by radical groups in Eastern Europe. An early document of the Solidarity trade union in Poland challenged the autocratic model of modern urban planning and urged a postmodern architecture to restore historic memory, charging that 'Reducing architecture to its utilitarian function is to remove its role as a means of social communication' (Portoghesi, 1982, p. 46). In the Soviet Union, self-styled postmodern artists have produced bold paintings critical of totalitarian power, dating from the dangerous times preceding *glasnost* when there was no artistic freedom to protect such subversion (Buck-Morss, 1991). In such ventures there is no evidence of Lyotard's (1984) scholastic relativism, but rather a committed, indeed a moral, task as part of a democratic project.

This conclusion itself is incapable of resolving any political question. It does, however, produce more even ground for that question to be addressed.

The Modern project and beyond

In speaking of democratic projects we have moved from a discussion of cultural artifacts, from the sheer facticity of form, to a consideration of cultural, indeed political, practice. Since one of the more hopeful sides of postmodernity is a criticism of the non-democratic cultural and political practices of modernism and modernization, it is with this point that we shall principally be concerned.

To follow Habermas, modernity implies the differentiation of autonomous realms.[3] The process of differentiation categorizes the world into inherently autonomous objects each with its own set of rules: nature is distinguished from humanity, culture from society, ethics from aesthetics, high culture from mass culture, secular from religious, civilized from savage, men from women and so on. Each realm enjoys its own self-contained, self-governed dehistoricized rationale. For example, aesthetics endow the 'art object' with an essence independent of its social or natural context. At the same time, differentiation also involves the placement of hierarchy. Within the aesthetic realm, for instance, aesthetic objects are distinguished from non-aesthetic objects and enjoy a

[3]This general account of the modern project is, of course, nuanced by a diversity we cannot discuss here.

superior status—as exemplified by the notions of high culture and low culture. Similar hierarchies are constructed around gender, for instance, or race. Modernity, then, commonly proceeds by naturalizing hierarchies and prioritizing in each realm the claims of the superior element, thus devaluing alternative modes of authority: abstraction is elevated over particularity, the Same over the Other.

The essences of a particular practice, such as aesthetic practice, are accessible to the modern individual subject, a 'monad-like container, within which things are felt which are then expressed by projection outwards' (Jameson, 1984a, p. 63). Creative individuals thus act independently of their social context and are capable of initiating avant-garde activity, 'setting out before the rest of society to conquer new territory, new states of consciousness and social order' (Jencks, 1986, p. 32). According to Lash (1990a) the modern rule of abstraction is the metaphysics of progression, the movement towards 'ought' rather than the recovery of historical values.

This modern rule provides a blueprint for the built environment and the practices of planning and architecture. Premodern design derived its codes from the contingencies of the local area; modern practice sought a self-referential and universal language for good design, abandoning reference to the particularities of social or natural setting. Modern planning thus turned its back on the existing urban fabric, imposing instead an ordered differentiation of space by means of systematic zoning. At the architectural scale, modernism is exemplified by the glass box or concrete cage (Relph, 1987), near-sculptural space isolated from its local context. Although early proponents of architectural modernism justified it with reference to a utopian text of *social* transformation (e.g., Le Corbusier, 1927), in practice its social function is a secondary concern, overshadowed by the creative individual's duty to experiment in the arts of form and to explore the pure and inherent qualities of light and materials (Lash, 1990b). In sum, the modern landscape carries a text of order and rationality, of disjuncture from the past and a progression towards abstract ideals, and in all of this it reveals a clear imprint of authority.

In his review of modern architecture, Habermas (1989a) makes a series of defenses which account for the failings that he acknowledges. He notes first the practical task of meeting the massive challenges of modernization; second the consistency of a functional aesthetic to trends both in art and in industrial society; and third the overly ambitious social program of the modern manifesto. There is, to be sure, substance to these arguments, but they exclude from any blame the logic of the modern project itself. This omission becomes most evident in his conclusion, where Habermas looks for some encouragement for a renewal of architectural modernism. He alights upon community architecture, which engages local ecology, historical preservation and 'a close connection between architectural design and spatial, cultural, and environmental contexts' (Habermas, 1989a, p. 19). In an uncharacteristic sleight of hand these quintessential traits of postmodern contextualism are appropriated as revealing 'something of the impulse of the Modern Movement,' although he concedes that such tendencies do display 'a veneer of antimodernism'! Once again we see how slim is the divide between Habermas and the (unmentionable) postmodern.

Returning to his endorsement of community architecture (cf. Prince Charles!), Habermas finds 'especially noteworthy' those instances where planning is characterized by dialogue between planners and clients. As noted at the beginning of this paper, the dialogical model is precisely the one that we would expect Habermas to endorse from his own premise of the ideal speech situation. But the *power relations of the modern movement in architecture and planning consistently rejected such a dialogical model.* Unlike the defenses Habermas proffers for the failure of modernism, this shortcoming is not external to the modern project itself but lies at its very core. The modern avant-garde is by definition elitist—indeed its survival as a critical force is warranted only as long as it avoids contamination from mass culture. In Marcuse's words: 'would not an art which rebels against integration into the market necessarily appear as 'elitist'?' (Marcuse, 1978, cited in Brantlinger, 1983, p. 232). The prophetic role is reassigned as art replaces religion; into Nietzsche's empty and desacralized space steps the artist as Superman.

The history of modern architecture has commonly shown that the frequent neglect of local context has been accompanied by a demeaning view of local users of space. Gropius found his working-class clients too 'intellectually undeveloped' to be consulted in his mass housing projects, while Le Corbusier also felt the necessity for people to be 're-educated' to interact meaningfully with his urban vision (Knox, 1987). In the Athens charter of the Congrès Internationaux d'Architecture Moderne (CIAM), Le Corbusier showed that for him the equivalence of the architect with Superman was not fanciful: the CIAM architect is the master social engineer because (s)he 'possesses a perfect knowledge of man' (cited in Holston, 1989, p. 77).

To an extent not shared by other modern artists, architects and planners were empowered to execute their top-down visions by the state and private corporations. So modernism became modernization, the application of CIAM principles to urban development across the globe. In a careful interpretation of the design and construction of Brasilia, Holston (1989) has detailed the remarkably direct extension of the CIAM manifesto to the urban landscape. In this process the client group is absent, for 'the only kind of agency modernism considers in the making of history is the intervention of the prince (state head) and the genius (architect-planner)' (Holston, 1989, p. 9). Indeed the title page of *The Radiant City,* Le Corbusier's most expansive urban vision, includes the note 'This work is dedicated to AUTHORITY' (Holston, 1989, p. 42). The total decontextualization of the plan for Brasilia was part of the avant-garde strategy of shock, of defamiliarization. But with Brasilia the shock was not simply aesthetic, but part of a grand strategy of social engineering to break down existing patterns of culture and everyday life in order to build a new society. Holston and many others have exposed the fallacies of this grand social vision on the ground. Here our concern is to emphasize the anti-democratic political processes that empower it.

In Brasilia modernism was intended to be a vehicle for the modernization of the whole country, and in North America the modernization of the urban fabric was similarly undertaken under the auspices of the modern movement. For Giedion (1967), the official historian of modern architecture and planning, the grand designs of Robert Moses, builder of New York's highways and public works for over a generation, represented the climax of modernity applied to the city. Moses, with his firm grasp of authority, had no time for the complications of participatory planning. For his urban 'meat ax' the problems were technical, not human: 'more houses in the way . . . more people in the way—that's all' (Berman, 1982, pp. 293–4).

It is important not to lose sight of CIAM's utopian ideal of a more egalitarian society, but as we have seen these ends were pursued by less than democratic means. By the 1950s there were already indications that the ends were not being achieved either. A younger generation of architects from within the CIAM fold noted that basic human needs of belonging, identity and neighborliness were not being met by the abstract protocols of functionalist design. In a remarkably prescient passage they wrote that 'the short narrow street of the slum succeeds where spacious redevelopment frequently fails' (Frampton, 1985, p. 271). But no solutions flowed from this insight. Either the group was cognitively unable to break with the conventions of modern design or else, in the case of Aldo Van Eyck, an anthropological critique of the functional city and its 'cultural void left by the loss of the vernacular' led to a deep pessimism (Frampton, 1985, p. 276). Rather than liberating the lifeworld, modern design was part of an invading system that was oppressing it—although (as Holston showed in Brasilia), residents commonly contested the oppressive spaces thrust upon them.

The recovery of the vernacular and attention to issues of place and identity required, among other things, a dialogical method involving other speech communities that would permit a defamiliarization or deconstruction of the assumptions of the modern project itself. Using the same prototype of 'the short narrow street of the slum' Jane Jacobs (1961) made a decisive advance not only in deconstructing the modern text of function and rationality, but also in re-opening the path to the vernacular in dialogical method by introducing other voices which interrogate and revise the epistemological and ontological categories of modern architect-planners and their enabling authorities. Focussing on the discourse of planning for Boston's inner neighborhoods, she identified two alternative possible bases to authority. One was grounded in the qualities of the neighborhoods themselves: ugly, vital, messy, complicated. The other, grounded in the academy, was the modern discourse of the professional, concerned with order, rationality, uniformity, and the necessity of urban renewal. Visiting the North End, Jacobs' planner friend is aware of both claims:

> *'I know how you feel,' he said. 'I often go down there myself just to walk around the streets and feel that wonderful, cheerful street life . . . But of course we have to rebuild it eventually. We've got to get those people off the streets.'*
> *Here was a curious thing. My friend's instincts told him the North End was a good place . . . But everything he had learned as a physical planner about what is good for people and good for city neighborhoods, everything that made him an expert, told him the North End had to be a bad place.*
>
> Jacobs, 1961, p. 20

From the point of view of disenfranchised clients, however, 'there is a quality even meaner than outright ugliness or disorder, and this meaner quality is the dishonest mask of pretended order, achieved by ignoring or suppressing the real order that is struggling to exist and to be served' (Jacobs, 1961, p.25).

Modern planning is justified by its utopian agenda, but in effect—and in isolation—it has often proven elitist and alienating. Here, then, we see the elevation of one discourse which seeks to impose a universal solution and the suppression of another discourse which is contextually founded, modest in its claims to a bounded authority for that time and that place. From this point of view, the modern perspective could be one among many, but it has been a privileged one due to its social and economic position and due to the status of the professional expert—indeed, due to precisely those contingent influences, the force of which its egalitarianism denies. Jacobs' deconstruction, however, throws into question modernist epistemology and ontology as well as the political hierarchy implicit in planning practice. She thereby opens up 'a field of tension between tradition and innovation, conservation and renewal, mass culture and high art, in which the second terms are no longer automatically privileged over the first' (Huyssen, 1984, p. 48).

But deconstruction, this 'critique of origins' (Foster, 1983, p. xii; Dear, 1986), is only half the story. Postmodernism is also reconstruction. Having dismantled the language that justifies modernism, it must be reassembled in some new and different format—but one that is always provisional and open in turn to deconstruction itself. For the example of postmodern architecture, the parts may not be new, and the whole is thus experienced less as 'a *work* in modernist terms—unique, symbolic, visionary—than as a text in a postmodernist sense—'already written,' allegorical, contingent' (Foster, 1983, p. x). But the practice of reassembling the parts is simultaneously familiar yet unpredictable.[4] The

[4]This strategy of montage was employed commonly in modern art and literature. Increasingly it appears that (continued...)

new form may be designed to force a confrontation between elements drawn from competing traditions and now thrust together; for instance, a building might combine elements from vernacular building traditions with elements of Classical architecture and it might include both premodern and modern components, both elite and popular codes. The defining quality of postmodern architecture, as specified by Jencks, is precisely this 'double coding,' which allows a building both to function effectively and to represent symbolically the languages of various audiences at once. There is no straightforward design solution to a particular project; it is, however, possible to use the contemporary techniques and materials while also addressing—perhaps re-presenting—local vernacular architecture, the natural geographical context and the codes of potential users. Whereas modern buildings set themselves apart from the city, postmodern buildings thus 'celebrate their insertion' into the urban fabric, 'thereby renouncing the high modernist claim to radical difference and innovation' (Jameson, 1984b, p. 64). Postmodernism is thus a sensitivity to existing difference rather than an impulse towards the totalization of the new. Postmodern cultural practice is addressed inward to a specialized speech community but equally it is directed outward to users who occupy a different social and cultural setting; when unitary notions of authority are challenged, then opportunities are opened up for differences to be accepted and promoted: a postmodernism of the Other (Turner, 1990).

It is, of course, very easy to itemize such desiderata, but very much more difficult both to conceptualize and accomplish successful solutions in the urban landscape. Undoubtedly critics are correct to identify any number of failures, but this is only to reinforce the fact that the achievement of 'a communicative, meaningful architecture,' the impasse modernism identified but could not address, remains a substantial conceptual problem in an age of relativism (Norberg-Schulz, 1988). Recognition of the problem of representation is no guarantor of an easy solution (Jencks, 1983; Ley, 1989). It bears repeating that the need for such an architecture is acknowledged both by modernists (like Habermas) and postmodernists (like Portoghesi). But as we have said before, a treatment of form alone is insufficient. Habermas (1989a), among others, has properly observed that we should not repeat the mistake of CIAM and expect more from architecture than it can achieve. Jacobs pointed to the broader question of planning, of an appropriate urban vision. Her criticism was directed, moreover, beyond built form and against a purely rational epistemology, a centralized politics, and a functionalist ontology of the undifferentiated public interest.

[4](...continued)
discontinuities between modernism and postmodernism are quite variable between artistic genres. Perhaps architecture is treated as the paradigmatic case of postmodernism in part because in architecture the discontinuity with what went before is exaggerated to an unusual degree.

Toward a postmodernism of resistance?

We have argued with others that the cultural elitism of the modern project, its belief in the superior insight of the artist/architect, not only produced a self-referential aesthetic, but also implied an asymmetric politics which disenfranchised popular participation. From this criticism, a central element of a postmodern aesthetic is a dialogical model when conversations between the artist and local cultures should sustain intersubjective senses of identity and place. But such an aesthetic also has its political counterpart. Multiple voices can only be heard when they are empowered, when they are located in non-hierarchical political spaces. Thus if the tendencies of the modern movement are toward a politics of exclusion, the tendencies of the postmodern agenda are toward a politics of inclusion. Of course, whether or not such tendencies are fulfilled or deflected is a separate issue.

These general considerations provide a context for reviewing the social protests against the modern city in the 1960s and early 1970s. The vast modernization of the urban fabric during this period—urban renewal, new highway systems, urban redevelopment—represented a concentrated program of rebuilding after the inactivity of the Depression and World War Two. As Jacobs (1971) has noted, the idiom of reconstruction was the 1920s vision of the freeway, high-rise city. Resistance to this program was a resistance to both form and process, to both the alienating style of modernism and the destructive effects of modernization upon the existing urban landscape. The postmodern theme of contextuality redirected attention to 'sensitive urban place-making' (Jencks, 1981, p. 82) and away from the 'abstract universalism' (Rustin, 1987, p. 31) of modern spaces. What Rustin calls 'a new particularism' has attempted over the past twenty years to re-establish the supportive integration of place and social identity (Robins, 1988). The animation of public space—where the models of the town square and 'the short narrow street of the slum' have commonly been prototypes—has included the celebration of difference in the ethnic and lifestyle festivals of a postmodern urban culture, often sponsored by leftist city councils (Bianchini, 1987; Robins and Gillespie, 1988). New styles of social housing, the reclamation of the street from its modernist dedication to traffic circulation alone and enhanced attention to design and landscaping have been part of a larger program of planning in the 1970s and 1980s. Public art of a critical postmodernism has sought to lay bare the concealed meanings and social consequences accompanying the production of urban space (Deutsche, 1988), for example, in the ironic criticism of consumer culture coded in the design projects of the SITE partnership, where 'comment is made on the soulless, exhausted, and ruined environment and indifferent nonarchitecture becomes striking antiarchitecture' (Fischer, 1985, p.261).

Our argument is not, of course, that these objectives have been easily or even widely attained. The task is rather to show that a postmodernism of resistance is neither a theoretical nor a historical mirage. Locality is invariably contested and new urban landscapes are always outcomes of a contingent political process. Moreover, protests against the modern city have been directed against more than the built environment. They have also been concerned with a politics of inclusion—the claims of visible minorities and other marginalized groups to cultural recognition and political enfranchisement in civic administrations dominated by white, middle-class businessmen. Leadership of urban protests has been frequently coordinated by a growing segment of the middle class, social and cultural professionals often employed in the public or non-profit sectors who felt equally alienated from existing power bases and were joined by pre-professionals, particularly students, expressing what Peter Berger has called the demodernizing consciousness of the counter-culture (Berger et al., 1973, p. 208; also Offe, 1987). The same cohort has reappeared in other social movements: environmentalism, cultural nationalism and civil rights, including some feminist groups and the democracy movement.[5] Here is the domain of a critical postmodern politics and each of these movements share important common themes in their resistance to modernization. Fundamental is a suspicion of the unconstrained power of both the state and the market. Calls for the diffusion of their influence and decentralization of their power are accompanied by a more plural set of societal goals and political voices than economic development and its attendant power elites. With modest revisions an alternative and critical social paradigm developed by environmentalists (Cotgrove and Duff, 1980) may be extended to some of these other domains. The core values of this paradigm recognize broader public interests than market forces or state power; they acknowledge the plurality of human and non-human rights in a complex society; they advocate more participatory and non-hierarchical political structures; they endorse small-scale and locally sensitive development and a more egalitarian society which is needs-driven; and they urge recognition of the limits of instrumental rationality and technical solutions. Because of their criticism of the postulates of modernism and modernization, such an alternative social agenda is said to be contained within a postmodern politics (Aronowitz, 1988; Betz, 1989; Mouffe, 1988).

This account, with its emphasis on the 1960s and 1970s, concurs with the periodization of both Huyssen (1984) and Jencks (1986), who see an initial critical positioning of the postmodern. But what happened after, say, 1975? Here they suggest

[5]We are not, of course, suggesting a simple homogeneity either between these movements or, indeed, within them. Nonetheless, their leaders frequently share a critique of modern society from the position of social and cultural professionals who are marginal to the dominant practices of the market and the state. More broadly this cohort features in theories of the new middle class. In a more detailed treatment it would also be necessary to examine more carefully their own position of authority within social movements; at one extreme, for example, some theorists of cultural nationalism see the advocacy role of the cultural new class as self-serving.

a stage of cooptation and commodification which becomes increasingly strong into the 1980s, a view not very different from the critical assessments discussed earlier. A postmodernism of reaction becomes the cultural expression of neoconservative politics. But there is another all-too-obvious possibility. Jameson has made the simple point that 'what matters in any defeat or success of a plan to transform the city is political power' (Stephanson, 1988, p. 15). The social movements of the 1960s successfully penetrated the state at different levels, and for a period adversarial politics seized an unprecedented share of the policy-making agenda. Not only was this period the high-water mark of the welfare state, but also a period of decentralist politics pursuing plural social objectives and showing successful signs of resistance in many cities to the alienations of continuing modernization in the image of modernism. For a variety of reasons the political coalitions of that period fell apart, and power recentralized. But this does not mean they are a spent force historically. In any case, the objective here is not to define a successful politics, but a *resistant* politics. Nor does it seem necessary to claim, as some authors do, that postmodern politics is a total politics completely redrawing past configurations. Clearly economic inequality will continue to be a major base of mobilization, though the claims to economic justice are now directed as much to state policy as to the labor contract. The potential for a new round of coalition-building remains; the populism of Jesse Jackson's Rainbow Coalition, for example, reached beyond conventional black politics to link a series of social movements, including environmentalism and feminism as well as civil rights. While electoral success remained concentrated in states with a large black population, primary victories in such unlikely places as Vermont and Alaska suggested that the coalition had broader support (Rogers, 1990). The radical capacity of such coalitions has been demonstrated most fully by the democracy movement in Eastern Europe, led by the cultural new class of professionals and intellectuals.

But the democracy movement raises a complication to this line of argument. Democracy does mean access to political enfranchisement and civil rights, but it also means access to full supermarket shelves. And so finally, and cautiously, we turn to the 'debased' postmodernism of reaction.

Here, it should first be noted that the equivalence drawn between postmodern cultural forms and middle class privilege does not bear the *necessary* causality ascribed to it. For example, it has been noted that during the past decade postmodern design has been directed to such middle class icons as country houses, art museums, ski chalets and festival markets. This does not, however, reveal a necessary causal link with neoconservatism, for particular appropriations during the past decade do not exhaust a broader role for postmodernism itself. Indeed, as we shall see, social housing over the past twenty years has usually turned from austere modernism to more contextual postmodern motifs, but because there is so much less of it in a privatized era it does not have the same emblematic power as the designer locales of middle-class culture.

A second reflection on the so-called postmodernism of reaction is no doubt more tendentious. The landscapes of mass culture (and it is mass, not popular, culture that lies behind this conceptualization) are described in terms of consumer spaces of spectacle and the disneyesque and then dismissed in the convenient label of commodification, an overworked term which needs to be demystified. We are suggesting a more nuanced point which encompasses a more complex interpretation of the market, where what is hidden from consideration in the name of commodification reappears to line the shelves of supermarkets and bookstores. It is impossible to be unaware of the alienations of the market; this is a valid criticism as far as it goes and must be sustained, but it remains significantly incomplete. What the democracy movement reinforces is a perception that the market simultaneously *empowers*, albeit, and inherently, unequally. A fundamental resistance is a resistance to scarcity, a fundamental struggle is the struggle for survival. Access to goods (as basic as bread) is as much a facet of democratization as free elections and guarantees for the rights of the marginalized. Thus imperfectly and unequally the Janus face of commodification reveals also democratization; the hardback editions for the few become the paperback editions for the many. To this degree criticism of the consumption rights of others needs to be cautiously (and self-critically) shaped, and care taken that indiscriminate use of such terms as 'disneyfication' is more than a statement of personal taste made possible by economic advantage (cf. Bourdieu, 1984). We suggest this conclusion may be a more general development of the point that Jameson is reaching for when he is pressed, but declines, to disqualify cultural postmodernism: 'Think of its popular character and the relative democratization involved in various postmodernist forms . . . Postmodern architecture is demonstrably a symptom of democratization' (Stephanson, 1988, p. 12).[6]

A concluding vignette

The final section of the paper moves away from the perilously modern enterprise of contextless argumentation. If the validity of a proposition depends on how it is situated, then our argument must be grounded in empirical events. One danger of treating postmodernism as an epoch is a neglect of how processes operate across space at different places and with differential effects. We need to contextualize the discussion geographically, precisely because the *achievement* of a landscape of resistance is only attainable subject to severe contingencies, associated, for instance, with the political context

[6]A different, but equally democratic, defense of elements of mass culture including film and architecture was made by Benjamin (1936).

operating in a specific locality. Our argument has not, of course, been conducted in a geographical vacuum. The interpretation has been guided by events in the cities we know best, those in metropolitan Canada. It is time now to make that regional point of reference more explicit.[7]

In urban Canada the stimulus for social protest in the late 1960s was often freeway construction or urban redevelopment, in resistance to the vision of the highway, high-rise city. In the City of Vancouver, for example, resistance erupted in the late 1960s to separate but linked projects of urban renewal and freeway construction that would have obliterated the Chinatown-Strathcona areas of retailing and settlement and the Gastown district, the original town site. In the latter case a coalition of arts and heritage groups successfully argued the case for the preservation of historical memory against a megaproject of private office redevelopment. In the former case, the long-ignored voices of Chinese-Canadians, marginalized in the dominant politics of Orientalism, were at last heard. In each case the *tabula rasa* of the urban bulldozer, its erasure of history, difference and the vernacular, was averted. But the protests transcended immediate land-use conflicts as a new politics of inclusion emerged. New civic political parties were founded in 1968, a liberal reform grouping of younger professionals and a socialist party with a strong anti-poverty agenda. Both objected to the process as well as the product of a generation of conservative, at-large municipal administrations led by small- business-owners with a single-minded orientation to economic growth. A postmodern politics, in short, was directed against the form of the modern city but also against its dominant social and political processes, particularly the style of technical, centralized control at City Hall sealed from community participation. In the 1970s the liberal party enjoyed electoral success; in the 1980s, the socialist party enjoyed periods in power.

However, empowerment for a postmodernism that went beyond resistance to reconstruction was more apparent in the earlier decade, when all three levels of government responded to the critical social movements of the times. After 1968 the new federal administration, led by a cultural professional, Pierre Trudeau, announced an 'open society,' which provided points of access to marginalized groups. Its urban and housing policies privileged rehabilitation over renewal and decentralized participation over centralized control. At the same time the first national initiatives in multiculturalism extended the citizenship rights of minorities in ways that have become increasingly palpable in the 1980s. The provincial government in British Columbia, 1972–75, was controlled for the first time by the New Democratic Party, a union-based social democratic party, which introduced a range of reforms including Community Resources Boards, in

[7]For a discussion of culture, politics and landscape in Vancouver which treats these themes more fully, see Ley, 1987, 1992; Ley and Olds, 1988; Mills, 1988.

which elected local committees administered and funded a range of social services. At the municipal level a reform council in the city of Vancouver introduced a participatory process of neighborhood land use planning.

There is not space here for a complete discussion of the politically inspired interventions in the built environment over the past twenty years, a litany of promises, compromises, frustrations and some successes. Because it is so germane to the argument of this chapter and includes consideration both of landscape form and of constitutive process, the example of cooperative housing is both relevant and hopeful as an illustration of a postmodernism resistant simultaneously to the inequalities of the housing market and the overbearing control of the state. Housing cooperatives are a form of social housing that require state subsidies, the contribution of share capital by residents (but without the status of an equity investment), their participation in design and self-management. During the 1960s discussion and mobilization developed around new forms of non-market housing in Canada, as elsewhere. It was prompted by a reaction against high-rise public housing in particular, housing for the masses that expressed all too well the agenda of the modern movement. A critical discourse resonated with such phrases as 'ghettos for the poor,' 'cell-like slabs' and 'centers of stigma.' What was at stake, however, was not merely the appearance of the structure, important though this was, but also the centralized process of construction and management that accompanied it. Community development workers urged greater participation and control by the residents of social housing to lessen dependency and stigma. Building upon several grass-roots experiments in cooperative housing, labor, cooperative and church groups founded in 1968 the Cooperative Housing Foundation of Canada (Selby and Wilson, 1988). The timing for this initiative could not have been better, for the new federal Liberal government under Pierre Trudeau established a Task Force on Housing and Urban Development the same year. Mindful of the extensive criticism of urban renewal and large-scale public housing and enabled by the social experimentation of the early Trudeau years, the Hellyer Task Force produced a sternly critical report of both the design and the process of housing construction for the poor. In 1970 the Cooperative Housing Foundation received a seed grant to build five projects; by 1988 there were over 52,000 cooperative housing units in (primarily metropolitan) Canada.

At the urban level a similar transformation of ideology occurred concerning social housing, particularly in Toronto and Vancouver, where reform councils were elected in 1972 on the heels of social protest against redevelopment and freeway proposals. Thereafter, in each city there was 'a clear philosophical rejection of 'modernist' approaches to urban design and architecture' in public projects (Hulchanski, 1984). The new face of social housing was particularly interesting in Vancouver, where the City Planning Department's requirement of both innovative design and 'neighborliness' encouraged postmodern architectural solutions. Co-op designs have sought to escape standardization, to break up massing with façade articulation, to build at a domestic scale with a 3–4 story maximum and to introduce local, vernacular elements with other features including porches, gables and chimneys, providing the textual discourse of 'home.' This postmodern landscape has attracted considerable approval: the 'rejection of post-war Modernist housing theories' has provoked a 'sense of urbanity . . . awareness of streetscape, a celebration of the collective, a sense of exuberance . . . over the past decade some of the best urban housing in Canada has come out of the social housing sector in Vancouver' (Berelowitz, 1988).

But as we have repeated throughout this chapter, an emphasis on form alone is insufficient. Behind the form lies a shared social project, expressed in communal space, joint management and participation with architects in the design process. Reflecting the ideals of the early 1970s, co-ops are socially mixed with an income range. It would be incorrect to present either the co-op model or urban planning more generally in Vancouver as utopian, for it is flawed like any other. We have presented its more hopeful face here to make the simple point that both an architecture and a politics of resistance to anti-democratic tendencies in the state and the market are possible. The gains are real if still far from adequate. Indeed to conservative political and economic tendencies co-ops are not merely marginal but subversive, and since 1986 the Conservative federal government has sought to reduce their role and redefine their status. A postmodernism of resistance must always be contested. But despite a literature to the contrary, our objective has been to show that such landscapes remain a possibility, theoretically and historically. Within the silences of an adversarial literature are spaces for resistance offering political opportunities which should not be dismissed out of hand.

Acknowledgment

The authors are grateful to Derek Gregory for a critical reading of this chapter.

References

AIA (American Institute of Architects) 1983 Postmodernism: Definition and Debate, *Journal of the American Institute of Architects* 72, 238–301.

Aronowitz, S. 1988 Postmodernism and Politics, in A. Ross (ed.), *Universal Abandon? The Politics of Postmodernism.* Minneapolis: University of Minnesota Press.

Benjamin, W. (1936) 1969 The Work of Art in the Age of Mechanical Reproduction, in H. Arendt (ed.), *Illuminations*. New York: Schocken.

Berelowitz, L. 1988 The Liveable City: Social Housing in Vancouver, *Canadian Architect*, February, 34–7.

Berger, P., Berger, B. and H. Kellner 1973 *The Homeless Mind: Modernization and Consciousness*. New York: Random House.

Berman, M. 1982 *All That Is Solid Melts Into Air: The Experience of Modernity*. New York: Simon and Schuster.

Betz, H.G. 1989 Postmodern Politics and the New Middle Class: The Case of West Germany, unpublished paper, Department of Political Science, Marquette University.

Bianchini, F. 1987 Cultural Policy and Changes in Urban Political Culture: The 'Postmodern Response' of the Left in Rome (1976–85) and London (1981–86), paper presented to the European Consortium for Political Research, Politics and Culture Workshop, Amsterdam, April.

Bourdieu, P. 1984 *Distinction*. London: Routledge & Kegan Paul.

Brantlinger, P. 1983 *Bread and Circuses: Theories of Mass Culture as Social Decay*. Ithaca, N.Y.: Cornell University Press,.

Buck-Morss, S. 1991 East-West: Is There a Common Postmodern Culture?. Paper presented to the series Art and Society, Vancouver Art Gallery, April.

Chorney, H. 1990 *City of Dreams: Social Theory and the Urban Experience*. Toronto: Nelson.

Cotgrove, S. and A. Duff 1980 Environmentalism, Middle Class Radicalism and Politics, *Sociological Review*, 28, 333–51.

Dear, M. 1986 Postmodernism and Planning, *Society and Space*, 4, 367–84.

Debord, G. 1973 *Society of the Spectacle*. Detroit: Black and Red.

Deutsche, R. 1988 *Uneven Development: Public Art in New York City*. October, 47, 3–52.

Deutsche, R. 1991 Boys Town, *Society and Space*, 9, 5–30.

Fischer, V. 1985 SITE, in H. Klotz (ed.), *Postmodern Visions*. New York: Abbeville Press.

Foster, H. 1983 Postmodernism: A Preface, in H. Foster (ed.), *The Anti-Aesthetic: Essays on Postmodern Culture*. Port Townsend, WA: Bay Press.

Foster, H. 1985 (Post)modern Polemics, in H. Foster (ed.), *Recodings: Art, Spectacle, Cultural Politics*. Port Townsend, WA: Bay Press.

Frampton, K. 1985 *Modern Architecture: A Critical History*. London: Thames and Hudson.

Frieden, B. and L. Sagalyn 1989 *Downtown, Inc*. Cambridge, Mass.: MIT Press.

Giedion, S. 1967 *Space, Time and Architecture.* Cambridge, MA.: Harvard University Press.

Habermas, J. 1983 Modernity: An Incomplete Project in H. Foster (ed.), *The Anti-Aesthetic.* Port Townsend, WA: Bay Press.

Habermas, J. 1989a Modern and Postmodern Architecture, in *The New Conservatism: Cultural Criticism and the Historians' Debate.* Cambridge, MA: MIT Press.

Habermas, J. 1989b Neoconservative Cultural Criticism in the United States and West Germany, in *The New Conservatism: Cultural Criticism and the Historians' Debate.* Cambridge, MA: MIT Press.

Harvey, D. 1989 *The Condition of Postmodernity.* Oxford: Blackwell.

Harvey, D. 1990 Between Space and Time: Reflections on the Geographical Imagination, *Annals, Association of American Geographers,* 80, 418–34.

Holston, J. 1989 *The Modernist City: An Anthropological Critique of Brasilia.* Chicago: University of Chicago Press.

Horkheimer, M. and T. Adorno 1972 The Culture Industry: Enlightenment as Mass Deception, *Dialectic of Enlightenment.* New York: Herder and Herder.

Hulchanski, D. 1984 St. Lawrence and False Creek. Papers, No. 10, School of Community and Regional Planning, University of British Columbia.

Hutcheon, L. 1986–87 The Politics of Postmodernism: Parody and History, *Cultural Critique,* 5, 179–207.

Huyssen, A. 1984 Mapping the Postmodern, *New German Critique,* 33, 5–52.

Jacobs, J. 1961 *The Life and Death of Great American Cities.* New York: Vintage.

Jacobs, J. 1971 City Limits, Ottawa: National Film Board 1971.

Jameson, F. 1984a Postmodernism, or the Cultural Logic of Late Capitalism, *New Left Review,* 146, 53–92.

Jameson, F. 1984b The Politics of Theory: Ideological Positions in the Postmodernism Debate, *New German Critique,* 33, 53–65.

Jencks, C. 1981 *The Language of Post-Modern Architecture.* New York: Rizzoli.

Jencks, C. 1983 The Perennial Architectural Debate, in C. Jencks (ed.), *Abstract Representation.* London: Architectural Design Profile, Academy Editions.

Jencks, C. 1986 *What Is Post-Modernism?* New York: St. Martin's Press.

Jencks, C. and **W. Chaitkin** 1982 *Architecture Today.* New York: Abrams.

Kern, S. 1983 *The Culture of Time and Space 1880–1918.* Cambridge, MA: Harvard University Press.

Knox, P. 1987 The Social Production of the Built Environment: Architects Architecture and the Post-Modern City, *Progress in Human Geography,* 11, 354–77.

Lash, S. 1990a Postmodernism as Humanism? Urban Space and Social Theory, in B. Turner (ed.), *Theories of Modernity and Postmodernity.* London: Sage.

Lash, S. 1990b Lessons from Leipzig, or Politics in the Semiotic Society. Paper presented to the Institute of British Geographers, Glasgow, January.

Le Corbusier 1927 *Towards a New Architecture.* London: John Radker.

Ley, D. 1987 Styles of the Times: Liberal and Neoconservative Landscapes in Inner Vancouver, 1968–1986, *Journal of Historical Geography*, 13, 40–56.

Ley, D. 1989 Modernism, Postmodernism and the Struggle for Place, in J. Agnew and J. Duncan (eds.), *The Power of Place.* Boston: Unwin Hyman.

Ley, D. 1992 Co-operative Housing as a Moral Landscape, in J. Duncan and D. Ley (eds.), *Re-presenting Cultural Geography.* London: Harper Collins.

Ley, D. and **K. Olds** 1988 Landscape as Spectacle: World's Fairs and the Culture of Heroic Consumption, *Society and Space*, 6, 191–212.

Lyotard, J. 1984 *The Postmodern Condition.* Minneapolis: University of Minnesota Press.

Marcuse, H. 1978 *The Aesthetic Dimension.* Boston: Beacon Press.

McLeod, M. 1989 Architecture and Politics in the Reagan Era: From Postmodernism to Deconstructivism, *Assemblage*, 8, 23–59.

Mills, C. 1988 'Life on the Upslope': The Postmodern Landscape of Gentrification, *Society and Space*, 6, 169–90.

Mouffe, C. 1988 Radical Democracy: Modern or Postmodern? in A. Ross (ed.), *Universal Abandon? The Politics of Postmodernism.* Minneapolis: University of Minnesota Press.

Norberg-Schulz, C. 1988 The Two Faces of Post-Modernism, *Architectural Design*, 58, no. 7/8, 11–15.

Offe, C. 1987 Challenging the Boundaries of Institutional Politics: Social Movements Since the 1960s, in C. Maier (ed.), *Changing Boundaries of the Political.* Cambridge: Cambridge University Press.

Preziosi, D. 1988 La Vi(ll)e en Rose: Reading Jameson Mapping Space, *Strategies*, 1 (1988), 82–99.

Portoghesi, P. 1982 *Postmodern: The Architecture of the Postindustrial Society.* New York: Rizzoli.

Relph, E. 1987 *The Modern Urban Landscape.* London: Croom Helm.

Robins, K. 1988 Reimagined Communities? European Image Spaces, Beyond Fordism. Unpublished paper, Centre for Urban and Regional Development Studies, University of Newcastle-Upon-Tyne.

Robins, K. and **A. Gillespie** 1988 Beyond Fordism? Place, Space and Hyperspace. Paper presented to the International Conference on Information, Technology and the New Meaning of Space, Frankfurt, May.

Rogers, A. 1990 Towards a Geography of the Rainbow Coalition, 1983–89, *Society and Space*, 8, 409–26.

Rorty, R. 1985 Habermas and Lyotard on Postmodernity, in R. Bernstein (ed.), *Habermas and Modernity*. Oxford: Polity.

Rustin, M. 1987 Place and Time in Socialist Theory, *Radical Philosophy*, 47, 30–36.

Selby, J. and A. Wilson 1988 Canada's Housing Cooperatives. Research Paper No. 3, Cooperative Housing Foundation of Canada, Ottawa.

Soja, E. 1989 *Postmodern Geographies*. London: Verso.

Stephanson, A. 1988 A Conversation with Fredric Jameson, in A. Ross (ed.), *Universal Abandon? The Politics of Postmodernism*. Minneapolis: University of Minnesota Press.

Turner, B. 1990 Periodization and Politics in the Postmodern, in B. Turner (ed.), *Theories of Modernity and Postmodernity*. London: Sage.

Wexler, P. 1990 Citizenship in the Semiotic Society, in B. Turner (ed.), *Theories of Modernity and Postmodernity*. London: Sage.

INDEX